Complex Systems Design & Management

Gauthier Fanmuy · Eric Goubault
Daniel Krob · François Stephan
Editors

Complex Systems Design & Management

Proceedings of the Seventh International Conference on Complex Systems Design & Management, CSD&M Paris 2016

CESAMES
Center of Excellence on Systems Architecture, Management, Economy and Strategy

Springer

Editors
Gauthier Fanmuy
Dassault Systèmes
Vélizy-Villacoublay
France

Eric Goubault
École Polytechnique
Palaiseau
France

Daniel Krob
CESAMES
Paris
France

François Stephan
IRT SystemX
Palaiseau
France

ISBN 978-3-319-49102-8 ISBN 978-3-319-49103-5 (eBook)
DOI 10.1007/978-3-319-49103-5

Library of Congress Control Number: 2016956186

Printed on acid-free paper

This Springer imprint is published by Springer Nature
The registered company is Springer International Publishing AG
The registered company address is: Gewerbestrasse 11, 6330 Cham, Switzerland

Preface

Introduction

This volume contains the proceedings of the Seventh International Conference on "Complex System Design & Management" (CSD&M 2016; see the conference website: http://www.2016.csdm.fr/ for more details).

The CSD&M 2016 conference was jointly organized during December 13–14, 2016 at the Chesnaie du Roy at Vincennes (France) by the two following founding partners:

1. The non-profit organization Center of Excellence on Systems Architecture, Management, Economy and Strategy (CESAMES),
2. The Ecole Polytechnique—ENSTA ParisTech—Télécom ParisTech—Dassault Aviation—DCNS—DGA—Thales "Engineering of Complex Systems" chair.

The conference benefited of the permanent support of many academic organizations such as Ecole Polytechnique, CentraleSupélec, ENSTA ParisTech and Télécom ParisTech which were deeply involved in its organization.

We also would like to thank the conference partners: Dassault Aviation, DCNS, Digiteo Labs, Direction Générale de l'Armement (DGA), Institut de Recherche Technologique (IRT) SystemX, MEGA International and Thales which were the main industrial and institutional sponsors of the conference.

We are also grateful to several non-profit organizations such as Association Francaise d'Ingénierie Systeme (AFIS) and International Council on Systems Engineering (INCOSE) which strongly supported our communication effort.

All these institutions also helped us a lot through their constant participation to the organizing committee during the one-year preparation of CSD&M 2016.

Many thanks therefore to all of them.

Why a CSD&M Conference?

Mastering complex systems requires an integrated understanding of industrial practices as well as sophisticated theoretical techniques and tools. This explains the creation of an annual *go-between* forum at European level (which did not existed yet) dedicated to both academic researchers and industrial actors working on complex industrial systems architecture and engineering. Facilitating their *meeting* was actually for us a *sine qua non* condition in order to nurture and develop in Europe the science of systems which is currently emerging.

The purpose of the "Complex Systems Design & Management" (CSD&M) conference is exactly to be such a forum, in order to become, in time, *the* European academic–industrial conference of reference in the field of complex industrial systems architecture and engineering, which is a quite ambitious objective. The last six CSD&M Paris conferences—which were all held the last trimester of the year from 2010 to 2015 in Paris—were the first steps in this direction. In 2015, there were almost 300 participants who came from 20 different countries which measures the growing success of the CSD&M conference.

Our Core Academic—Industrial Dimension

To make the CSD&M conference this convergence point of the academic and industrial communities in complex industrial systems, we based our organization on a principle of *complete parity* between academics and industrialists (see the conference organization sections in the next pages). This principle was first implemented as follows:

- the program committee is composed of 50 % academics and 50 % industrialists,
- the invited speakers came in a balanced way from numerous professional environments.

The set of activities of the conference followed the same principle. They indeed consist of a mixture of research seminars and experience sharing, academic articles and industrial presentations, software and training offers presentations, etc. The conference topics cover in the same way the most recent trends in the emerging field of complex systems sciences and practices from an industrial and academic perspective, including the main industrial domains (aeronautic & aerospace, transportation & systems, defense & security, electronics & robotics, energy & environment, healthcare & welfare services, media & communications, software & e-services), scientific and technical topics (systems fundamentals, systems architecture & engineering, systems metrics & quality, systemic tools) and system types (transportation systems, embedded systems, software & information systems, systems of systems, artificial ecosystems).

The 2016 Edition

The CSD&M Paris 2016 edition received 46 submitted papers, out of which the program committee selected 16 regular papers to be published in the conference proceedings. A 29 % acceptance ratio was reached which guarantees the high quality of the presentations. The program committee also selected 17 papers for a collective presentation during the poster workshop of the conference.

Each submission was assigned to at least two program committee members, who carefully reviewed the papers, in many cases with the help of external referees. These reviews were discussed by the program committee during an online meeting by the May 30, 2016 and via the EasyChair conference management system.

We also chose nine outstanding speakers with various industrial and scientific expertise who gave a series of invited talks covering all the spectrum of the conference during the two days of CSD&M Paris 2016. The conference was organized around a common topic: *Challenges & Opportunities of Systems Engineering in a Changing World*. Each day proposed mix invited keynote speakers presentations and a "la carte" program comprising accepted papers presentations and conference partners' workshops.

Furthermore, we had a poster workshop, for encouraging presentation and discussion on interesting but "not-yet-polished" ideas. CSD&M Paris 2016 also offered booths and presentations to provide each participant a good vision of the latest engineering and technological news.

Paris, France
August 2016

Gauthier Fanmuy
Eric Goubault
Daniel Krob
François Stephan

Conference Organization

Conference Chairs

General Chair
Daniel Krob, Incose Fellow, CESAMES and Ecole Polytechnique, France

Organizing Committee Chair
François Stephan, IRT SystemX, France

Program Committee Co-Chairs
Gauthier Fanmuy, Dassault Systemes, France (industrial co-chair)
Eric Goubault, Ecole Polytechnique, France (academic co-chair)

Program Committee

The program committee consists of 21 members (10 academic and 11 industrial) of high international visibility. Their expertise spectrum covers all of the conference topics.

Academic Members

Co-Chair
Eric Goubault, Ecole Polytechnique, France

Members
Aleida Aleti, Monash University, Australia
Eric Bonjour, ENSGSI, France
Thao Dang, Verimag, France
Mike Hinchey, University of Limerick, Ireland
Daisuke Ishii, Fukui University, Japan

Claire Pagetti, Onera, France
Antoine Rauzy, Norwegian University of Science and Technology, France
Donna Rhodes, MIT, USA
Rafael Wisniewski, Aalborg University, Denmark

Industrial Members

Co-Chair
Gauthier Fanmuy, Dassault Systemes, France

Members
Sylvain Chabroux, LGM, France
Alain Dauron, Renault, France
Jeremy Dick, SyntheSys, UK
Jean-Paul Fardel, Airbus, France
Ali Koudri, IRT SystemX, France
Pascal Lamothe, PSA, France
Nicolas Gueit, Snecma, France
Garry Roedler, LMCO, France
Sven-Olaf Schulze, Unity AG, Germany
Lucio Tirone, Aster S.p.A., Italy

Organizing Committee

The organizing committee consists of 18 members (academic and industrial) of high international visibility. The organizing committee is in charge of defining the program of the conference, identifying keynotes speakers and has to ensure the functioning of the event (sponsoring and communication).

Chair
François Stephan, IRT SystemX, France

Members
Emmanuel Arbaretier, Airbus, France
Philippe Bourguignon, Engie, France
Alain Carof, DCNS, France
Etienne De Pommery, IRT SystemX, France
Johan D'Hose, Systematic Paris Region, France
Eric Duceau, Airbus, France
Brigitte Dueme, INRIA, France
Didier Dumur, Centralesupelec, France
Pascal Foix, Thales, France
Bruno Foyer, IRT SystemX, France
Vincent Gauthier, Telecom ParisTech, France

Omar Hammami, ENSTA, France
Michel Pinget, Dassault Aviation, France
Pascal Poisson, Alstom Transport, France
Garry Roedler, INCOSE, USA
Alain Roset, La Poste, France
Sylvie Tonda Goldstein, Ecole Polytechnique, France

Invited Speakers

Thierry Brizard, Executive Vice President Technology, CGG, France
Paul Eremenko, Chief Technology Officer, Airbus Group, France and Germany
Alan D. Harding, INCOSE President, UK
Paulien Herder, Head of the Engineering Systems and Services Department at the Faculty of Technology, Policy and Management, Delft University of Technology, Netherlands
Thierry Jean-Marius, Head of Systems Engineering Department, Airbus Safran Launchers, France
Virginie Maillard, Vice President Research, Renault, France
Garry Roedler, Incose Fellow and Engineering Outreach Program Manager, Lockheed Martin Corporation, USA
Matthew Silver, CEO and Founder, Cambrian Innovation, USA
Henri Verdier, Interministerial Director of the Digital Technology and the Information and Communication System of the French government (DINSIC), France

Acknowledgements

We would like to thank all members of the program and organizing Committees for their time, effort, and contributions to make CSD&M Paris 2016 a top-quality conference. Special thanks addressed to the CESAMES non-profit organization team which managed permanently with huge efficiency all the administration, logistics and communication of the CSD&M Paris 2016 conference (see http://www.cesames.net/).

The organizers of the conference are also greatly grateful to the following sponsors and partners without whom the CSD&M Paris 2016 event would not exist:

- **Founding Partners**
  - CESAMES—Center of Excellence on Systems Architecture, Management, Economy and Strategy,
  - Ecole Polytechnique—ENSTA ParisTech—Télécom ParisTech—Dassault Aviation—DCNS—DGA—Thales "Engineering of Complex Systems" Chair.
- **Academic Sponsors**
  - Ecole Polytechnique,
  - CentraleSupélec,
  - ENSTA ParisTech,
  - Télécom ParisTech.
- **Industrial and Institutional Sponsors**
  - Airbus Apsys,
  - Dassault Aviation,
  - DCNS,
  - Digiteo labs,
  - Direction Générale de l'Armement (DGA),
  - Institut de Recherche Technologique IRT SystemX,
  - MEGA International,

- PPI,
- Thales.

- **Supporting Partners**
  - Association Française d'Ingénierie Système (AFIS),
  - International Council on Systems Engineering (INCOSE).
- **Participating Partners**
  - Anylogic,
  - CCES—MIT,
  - IBM Analytics,
  - No Magic Europe,
  - Obeo,
  - PragmaDev,
  - The CoSMo Company.

Contents

Part I
Regular Papers

Challenges for MBSE and PLE for Legacy Product-Based System Environments

Michael Schäfer, Friedemann Bitsch, Stephan Weißleder and Florian Wartenberg

Abstract Model-Based System Engineering (MBSE) and Product Line Engineering (PLE) are well-known approaches in industry for the management and design of the architecture of complex systems. The railway signalling business has some specific characteristics that need to be considered in system engineering: railway signalling systems have a long life time and new systems have to integrate interfaces to many types of legacy railway safety products. This situation has led to different technical system approaches: railway infrastructure companies as customers prefer either turn-key projects fulfilled by one supplier or tend to define individual subsystems that can be integrated to a complete system. This article shows how Thales masters both approaches by using the method ARCADIA and the open source modelling tool Capella in the specific case of pre-existing subsystems and how the resulting variability is handled. An outlook will be given to extensions that allow an early safety analysis of models and will provide support for automatic test design.

M. Schäfer (✉) · F. Bitsch
Thales Deutschland, Geschäftsbereich Transportation Systems, Thalesplatz 1, 71254 Ditzingen, Germany
e-mail: michael.schaefer@thalesgroup.com

F. Bitsch
e-mail: friedemann.bitsch@thalesgroup.com

S. Weißleder · F. Wartenberg
Thales Deutschland, Geschäftsbereich Transportation Systems, Schützenstraße 25, 10117 Berlin, Germany
e-mail: stephan.weissleder@thalesgroup.com

F. Wartenberg
e-mail: florian.wartenberg@thalesgroup.com

G. Fanmuy et al. (eds.), *Complex Systems Design & Management*,
DOI 10.1007/978-3-319-49103-5_1

1 Introduction

1.1 History of Railway Signalling Systems

Railway signalling systems have been developed for more than 150 years to ensure the safe movement of trains [1, 2], offering the following basic functionalities:

(a) Provide a safe running path for each train in the railway network, avoiding collisions between different trains.
(b) Ensure that the train speed does not exceed a specific speed limit and especially that a train comes to stop in front of a signal at danger. This functionality avoids derailments due to excessive speed as well as collisions due to signals passed at danger.

In the past, separate systems have been developed for these basic functions: For providing a safe running path the so-called "interlocking" was invented. In the last 150 years the realizing technology has evolved from mechanical via electrical to software based systems, but the basic signalling and interlocking principles have remained the same.

Systems that control the correct movement of a train are called train control systems. Accidents, where trains wrongly passed a signal showing a danger aspect, have led to country dependent control systems that warn the train driver or automatically stop the train in such a situation. A unique European solution, the European Train Control System ETCS [3] has been developed over the last 20 years.

This historical evolution leads to a complex legacy environment for any new railway signalling application. New applications and systems have to be compatible with the installed base of signalling systems. Typical examples are:

- New computer based interlocking systems have to provide interfaces to a neighbouring railway station equipped with a mechanical interlocking built in 1900.
- New ETCS systems need to be combined with existing relay interlocking systems built in the 1950s.

Beside these legacy problems a manufacturer is confronted with two other issues: On the one hand all legacy or new interfaces differ from country to country. Even if they are generally standardized (e.g. ETCS) the required functionality is different for each country. So systems need to be adapted for every railway infrastructure company. On the other hand, systems grow more and more together. In the past interlocking train control systems were operated by different staff and have been loosely coupled. To save expenses by reducing the number of staff, systems get more and more coupled to automate and integrate operation.

1.2 *Different Architectural Approaches*

Faced with these problems, railway infrastructure operating companies like DB Netz, NetworkRail or SNCF Réseau follow different approaches:

One type of infrastructure company selects a complete renewal of all components of the railway signalling system all at once. An example is the Danish re-signalling programme [4]: From 2017 to 2021 the complete signalling system in the whole country will be replaced by an ETCS based system. A huge amount of money and the willingness to adapt long-term grown operational procedures to the new system concept are basic preconditions for this approach.

In comparison to this revolutionary approach, other operating companies agreed on a common architectural model specified in the European initiative EuLynx [5]. This architecture can be regarded as an extension of ETCS towards interlocking systems. But because of the different operational concepts the operating companies agreed on an interface model only: The detailed functions of each subsystem may differ from railway operator to railway operator. This evolutionary approach has the advantage that an upgrade of the railway system can be done step-by-step migrating to the EuLynx architecture. But it takes time and does not solve the problem of the large heterogeneity of the installed base.

2 Existing Approaches in System Engineering

Confronted with the diversity in the landscape of legacy systems described above, Thales selected in the past classical system engineering to manage the complexity of the systems. This chapter describes these approaches and the resulting consequences.

2.1 *Requirements Management*

Thales decided first to apply the methods of classical requirements engineering to cover the complexity. Customer requirements were transferred to system requirements and were distributed as requirements to the different subsystems:

This method ensured that no customer requirements defined by the railway operating company were missing. It also provided a good basis for testing the overall system as well as the different subsystems. But as Fig. 1 shows, this process has some disadvantages:

(a) Customer requirements are typically non-homogenous. In some areas customers are very experienced, so their requirements tend to be detailed. These requirements can be allocated nearly directly to a subsystem (shown as magenta in Fig. 1). In other areas (esp. the newer ones like ETCS) you may receive as supplier only very rough and fuzzy requirements that need to be

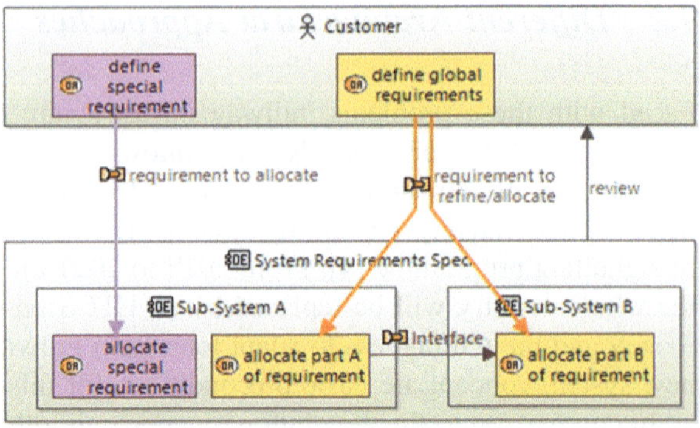

Fig. 1 Refinement of requirements

detailed and interpreted before they can be processed further. Concerning the legacy systems to be interfaced both variants exists: Some customers describe them in detail, while others may only provide a single requirement that a specific legacy system needs to be adapted.

(b) The refinement process is not only a requirements management process but also becomes an architectural process. Requirements are allocated to subsystems as functions and interfaces between the subsystems are defined, which are typically architectural tasks and not requirements management tasks (shown in orange in Fig. 1).

(c) Because the system requirements form the interface towards the customer, every detail that needs to be discussed with the customer is contained in the system requirements specification. This leads to an explosion of the number of requirements objects. The following list shows some examples:

- North-South Railway Saudi-Arabia: 2,730 valid requirements
- Danish re-signalling Programme: 5,600 valid requirements

It is obvious that this number can only be handled by persons that are deeply involved in a project. For newcomers these documents are nearly unreadable.

(d) Every system requirements specification is customer specific. A product line management approach was not applied, because each project organisation used to run requirement engineering to issue their perception of what the system of interest should do in support of the expected operational capabilities of the single customer. This approach hinders reuse: the same requirements could be interpreted differently, the same property can be described by different requirements sets. Functionality existing in different variants is not easy transferable, because already the system requirements specifications differ in form and content. This is a major problem concerning the legacy environment: A legacy interface realized for specific customer could not be easily transferred to a different customer, not to mention the case that this interface needs to be adapted.

Several approaches have been implemented to overcome this situation. One promising approach was to use Use-Cases in a textual form instead of classical requirements. This approach leads to better understanding and structuring of customer requirements, but it does not solve the problem that architectural tasks were done with the wrong methodology of requirements management.

2.2 *Modelling in UML/SysML*

In parallel to the approach of requirements management, modelling of system architectures with UML or SysML was started. These architectural drawings focus on a reverse engineering of the existing system architecture. They show the logical or the physical structure of the overall system and help people (from supplier and customer) to get a better understanding of how all the subsystems already mentioned in the System Requirements Specification are combined together.

But in combination with the classical requirements process listed above, several issues came up, that led to the fact that the modelling was regarded only as an additional task:

(a) Because the functionality was already defined and allocated in the requirements specification, the system architectural models and figures contain only boxes describing the system components and the (physical) interfaces but not their functionality. Often the tools are only used as drawing tools without relying on the specific advantages of a model in the background. An example is given in Fig. 2, which shows the high-level logical component structure of the system, the interfaces between the components and the surrounding actors.

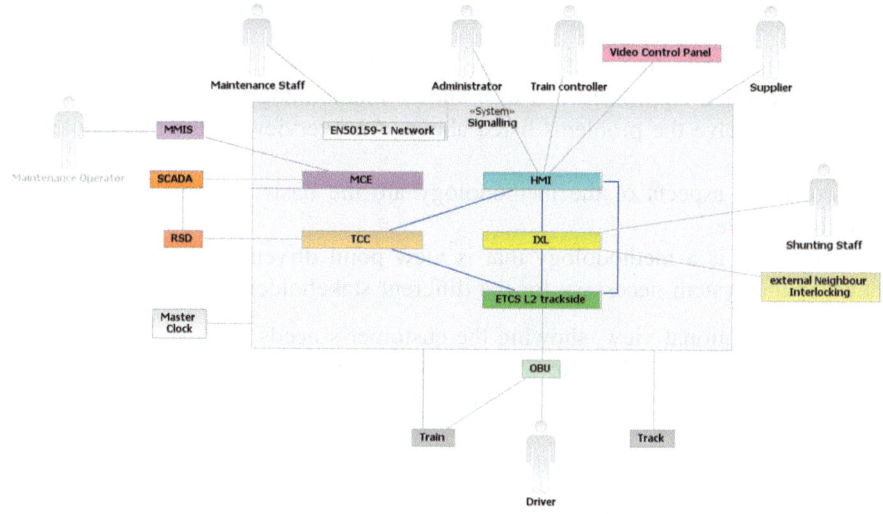

Fig. 2 UML component diagram of a signalling system

(b) UML and SysML are languages for modelling, they provide many different views, but they do not provide a methodology that helps system architects to know what to model in a diagram. This led even more so than in the area of requirements management to the problem that models and diagrams differ from project to project.
(c) The project-centric view results often in the case that legacy systems, even if they contribute largely to the system functionality, have been modelled only as external actors. This results in the problem that the next project, which needs this functionality (e.g. in a new implementation) could not reuse the model.

2.3 *Results of Classical System Engineering*

Classical System Engineering helped Thales to develop and deliver new railway signalling systems also in large and complex legacy environments. But it demonstrates also the limits of this methodology:

Handling of requirements specifications and system models becomes complex and heterogeneous. Legacy systems could be integrated, but the whole approach shows no concept for reusability.

3 Model-Based System Engineering and Product Line Approach

3.1 *ARCADIA and Capella as a Basis*

The system architecture methodology ARCADIA (ARChitecture Analysis and Design Integrated Approach) is a well-known methodology that empowers system engineers to solve the problems listed above. An overview of the methodology is given in [6].

Two major aspects of the methodology are the basis to overcome the issues described above:

ARCADIA is a methodology that is view point driven. It provides four basic views of the system necessary for the different stakeholders as shown in Fig. 3:

(a) The operational view, showing the customer's needs
(b) The system view showing what the system should provide
(c) The logical architecture showing how the subsystems provide this functionality
(d) The physical architecture showing the implementation of the functionality

ARCADIA is a "supplier-oriented" method in contrast to "customer-oriented" methods like "Zachman" or "NAF", which focus more on users capability,

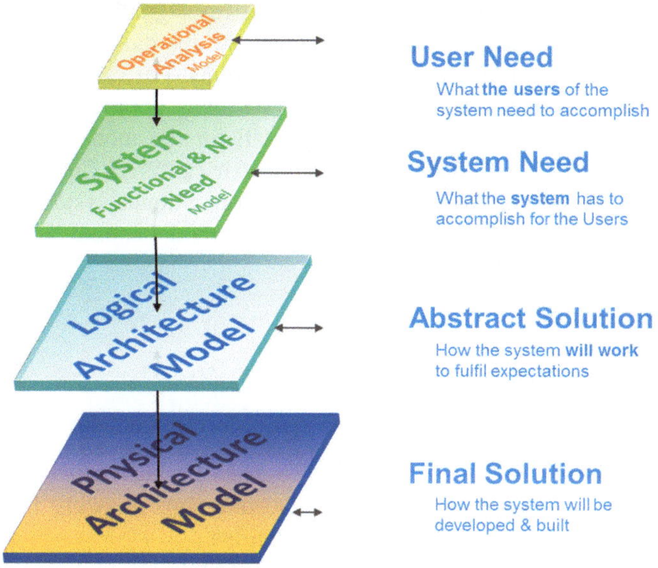

Fig. 3 Views of ARCADIA

acquisition and deployment. ARCADIA targets detailed solution definition and assessment:

- separating need and solution by applying different views,
- supporting stakeholders collaboration,
- dealing with complexity management,
- architecture evaluation, enforcing "correct by construction" modelling

Detailed Information about ARCADIA and the associated meta-model is provided in [7].

ARCADIA is a functional driven methodology. ARCADIA integrates the functional aspects originally covered by the requirements management directly with architectural aspects of subsystem allocation and implementation.

Capella is an Eclipse-based open source tool implementing the methodology of ARCADIA [8]. It was originally developed by Thales as an internal tool called Melody, so both names Capella and Melody are sometimes used as synonyms.

3.2 Reference System Architecture for High-Level PLE

ARCADIA as a method and Melody/Capella as a tool have been already applied in railway signalling business [9]. But this approach shows also that the methodology needs to be supplemented:

A specialisation of the methodology with railway signalling in mind was necessary. Although ARCADIA is a good framework for the system architecture it provides a lot of variants for the different viewpoints. For railway signalling systems, adequate modelling means were defined based on the available set. How this could be achieved is explained in detail in [10] for the model of the signalling system for the Passenger Rail Agency of South Africa "West Cape Region" (RSA project). The major result was that for each view (operational, system need analysis, logical architecture, and physical architecture), the model should be separated into three different but interrelated areas:

- Static architecture.
- Functional split
- Behaviour specification

The static architecture contains the structural description of the system. On the different views of ARCADIA this results in:

- Operational view: Operational architecture diagrams showing the operational entities in the system context and their relations.
- System view: System context diagrams illustrating the external system relationships and defining the outside border of the system.
- Logical view: Architectural diagrams describing the logical structure of the subsystems including their interconnections.
- Physical view: Architectural diagrams including the realization of the logical subsystems using software and hardware components.

The application of the functional split is an essential change to existing approaches. The system functions defined in the system view of ARCADIA are distributed on the subsystems of the static architecture. This functional split helps the architects to find a meaningful division into subsystems. So both areas are linked together as shown in Fig. 4 by the red arrows.

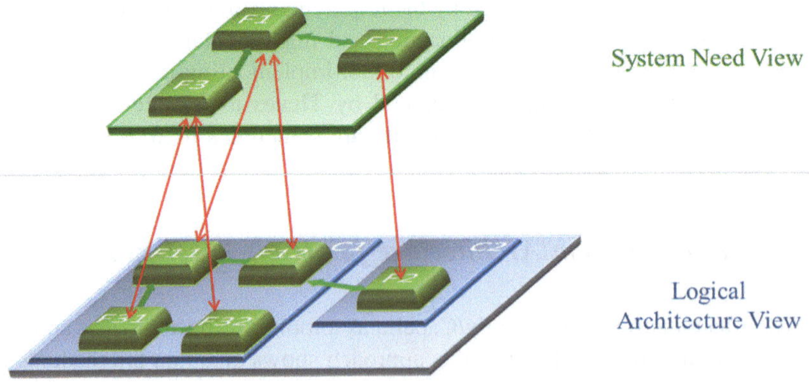

Fig. 4 Functional split in ARCADIA

The behaviour specification is the description of the detailed behaviour of each function.

Looking at the legacy environment described above in Chap. 2 it is essential to have this environment in mind. The borders and the functionality of existing systems have to be modelled in order to use the approach in legacy environments. But it is necessary not only to model the existing architecture, but also to keep a meaningful sub-division in mind. This leads to the fact that sometimes system functions need to be split into more logical functions than originally intended to provide a possibility to model legacy subsystems as well as new subsystems, too.

This approach leads to the next step, the introduction of variants to achieve reusability. Product Line Engineering with feature based variants is a well-established method to handle variability in software engineering [11]. The product line extension of the architecture of the Thales railway signalling system is called Reference System Architecture. Based on the modelling rules defined above a model is created that defines a standard architecture for a railway signalling system consisting of interlocking and train control systems. To enable Product Line Engineering this Reference System Architecture contains high level variants: Depending on a feature selection, either a legacy interlocking system or an integrated interlocking could be selected, as shown in Fig. 5 for a simplified example of route and signal handling.

Figure 5 shows that on the logical architecture level due to appropriate dimensioning of the subsystems no variant exists. Only on the physical level do different variants exist that can be selected during composition of a specific solution. It has to be stated clearly that not all possible variants are handled on this high system level. Detailed behavioural variants e.g. of interlocking logic or in the calculation of a movement authority in the train control system will be handled by (software) product line engineering on subsystem level. This approach helps to manage and reduce complexity and heterogeneity in way that they are only visible when needed.

4 Future Work

4.1 Model-Based Safety Analyses

Currently Thales is working on an approach to extend the ARCADIA methodology for safety analysis techniques. Safety analysis shall be based on the models of the System Architectural Design and the System Definition. It has to be avoided that safety engineers develop implicitly own system definitions for their safety analysis e.g. for the Technical Safety Report or for the architecture of Fault Trees. Therefore in our approach the fault tree analysis is based on the models in Capella. For that reason the model is enriched with corresponding failures and possibly also with failure rates (failure injection).

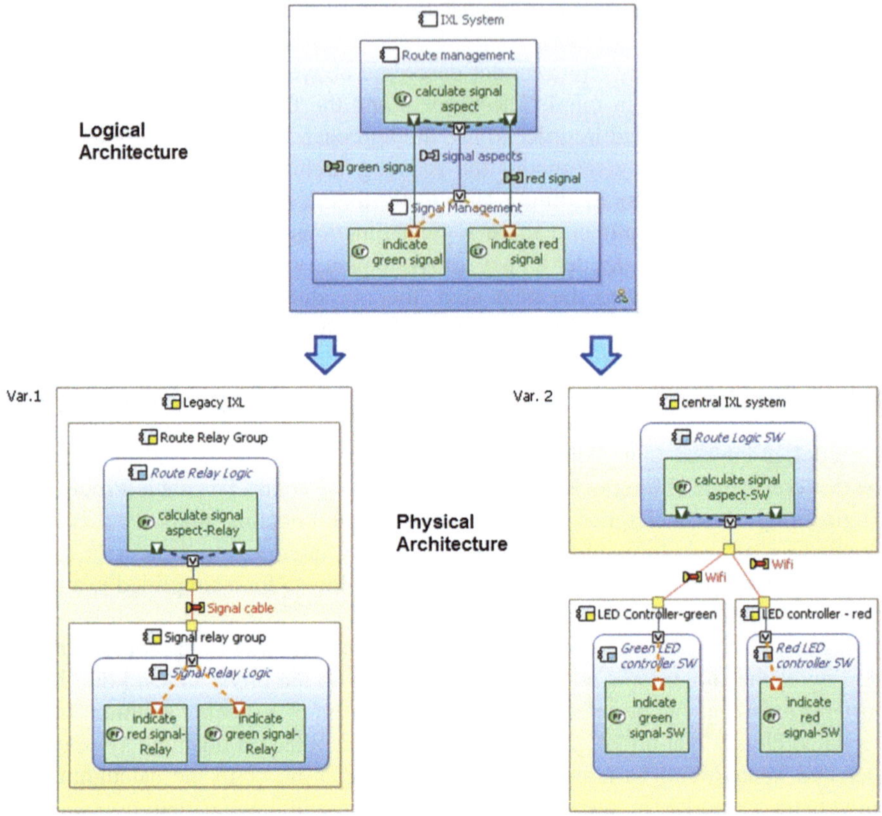

Fig. 5 Variants in railway signalling reference architecture

In order to identify errors in the manual Fault Tree analysis a fault tree is also automatically derived from the model and the result with the minimal cut sets can be used for the verification of the manual creation of the Fault Tree. On the basis of the derived fault tree structure and the specified failure rates in the model hazard rates can be calculated automatically.

For that purpose relevant parts of the Capella model are transformed into a formal framework, which is manually supplemented to a complete formal model by specifying the comprehensive internal behaviour of components in a formal language.

This formal model can then be also used for formal verification of safety requirements by model checking. Different model checking techniques shall be combined so that the most effective technique for the respective model structure can be chosen. In this way the correctness of the system definition in relation to the safety requirements can be shown.

As a consequence it is ensured that the safety case has the same basis as the system definition used in system engineering.

4.2 Automated Test Design

Here, we give a short outlook on how automated test design methods are introduced to significantly improve efficiency and effectiveness of test design.

Like many other engineering disciplines, test design is often a manual task: Test engineers read and understand requirements, derive test conditions, and create corresponding test cases. In a second step, validation engineers and system architects review the test cases in order to check that the test cases correspond to the linked requirements. The reason for this check is that this is an error prone task. Two of the most important sources of such errors are contradicting and incomplete requirements or different interpretations of requirements. This is mostly caused by the fact that requirements are written as plain text and leave room for interpretation. Further reasons may be an error in one of the described subsequent steps.

Automated test design based on models helps to avoid these issues: First, requirements are no longer described as plain text, but as models. This formal description of structure, behaviour, and their relations allows a holistic view on system requirements that minimizes the chance of misunderstandings, incompleteness, or contradictions: One can see relations in pictures instead of collecting information from several requirements. Secondly, a formal description of the system behaviour allows for automatically deriving test cases that check the correct implementation of this behaviour. This automation of the design process is fast, reproducible, and can incorporate subsequent changes in the model with significantly less effort (Fig. 6).

In the following, we describe the intended test generation process in detail. The process consists of steps on three abstraction layers: The data abstraction layer, the system behaviour layer, and the product line layer.

The data abstraction layer is the lowest one. The developed systems are data-driven systems. Hence the test cases contain much site-specific data. The motivation for this layer is the data-independent definition of test cases to ensure better reusability. Currently if there is a change in the data specification or the need to apply the same test cases to a different area, a significant effort for test design adaptation becomes necessary. In our approach, we define the test cases in a data-independent way and provide a test generator that maps these test cases to the

Fig. 6 Testing layers

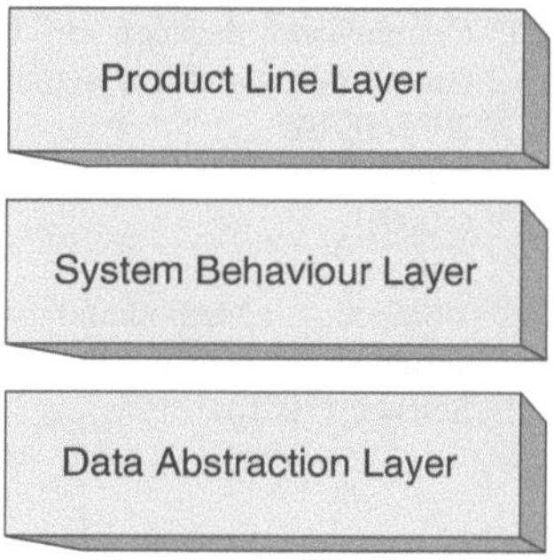

concrete site information. As a result, the test cases always fit the used station. To the best of our knowledge, this approach has not been applied before.

The system behaviour layer contains the step of deriving test cases from the system behaviour models. The resulting test cases can be produced in any given language. This approach of automated test design based on behavioural models is widely known and applied already [11–13]. In our case, we plan to generate test cases in the format of the above described data abstraction layer.

The product line layer covers the configuration of products in a product line. Feature models are typically used to describe this configurability. They can be linked to system behaviour models and can also be used to configure the behaviour models. We have invented methods to deal with this challenge [14] and plan to apply this to the railway domain.

5 Conclusion

This article shows, founded on the experience of classical requirements based system engineering, how a new approach to define the architecture of complex railway signalling systems has been developed. Based on the ARCADIA modelling methodology and the principles of product line engineering the approach provides a unique methodology that can help customers to state requirements at the right level and system engineers to define and implement a reusable Reference System Architecture. It covers variants (e.g. legacy systems) and provides a basis for future extension of model based safety analysis and automated test case generation. So it provides a means for controlling the evolution and migration of systems without diving into thousands of requirements allocated to hundreds of components.

References

1. Theeg, G., Vlasenko, S. (eds.): Railway Signalling and Interlocking—International Compendium. Eurailpress (2009)
2. Pachl, J.: Railway Operation and Control, 3rd edn. VTD Rail Publishing, Mountlake Terrace (USA) (2014)
3. Commission decision on the technical specification for interoperability relating to the control-command and signalling subsystems of the trans-European rail system, 2012-01-25. 2012/88/EU
4. Barfoed, L.: Current status of public transport in Denmark. Eurotransport, Issue **3**, 14–17 (2009)
5. Eulynx European Initiative. http://eulynx.eu. Accessed 25 Feb 2016
6. Voirin, J.-L.: Method and tools to secure and support collaborative architecting of constrained systems. In: ICAS 2010, 27th Congress of the International Council of the Aeronautical Science (2010)

7. Voirin, J.-L.: Modelling languages for functional analysis put to the test of real life. In: Proceedings of the Third International Conference on Complex System Design and Management CSD&M 2012. Springer (2013)
8. Polarsys.org: https://www.polarsys.org/capella/. Accessed 11 April 2016
9. Hoppen, A., Schäfer, M.: Modellbasierte Systementwicklung im Rahmen des Projects Denmark F-bane West, Eurailpress, Signal + Draht 09/2014, pp. 12–17
10. Müller, F., Schäfer, M.: Model-based system specification in the framework of the RSA project, Eurailpress, Signal + Draht 03/2016, pp. 31–36
11. Zander, J., Schieferdecker, I., Mosterman, P.J.: Model-Based Testing for Embedded Systems. CRC Press (2011)
12. Utting, M., Legeard, B.: Practical Model-Based Testing. Morgan Kaufmann (2007)
13. Weißleder, S.: Test Models and Coverage Criteria for Automatic Model-Based Test Generation with UML State Machines, PhD Thesis (2009)
14. Weißleder, S., Wartenberg, F., Lackner, H.: Automated test design for boundaries of product line variants. In: International Conference on Testing Software and Systems (ICTSS), 16 pages, Dubai, V.A.E. (2015)

The Trans-Alaska Pipeline System: A Systems Engineering Case Study

Robert S. Swarz

Abstract The Trans-Alaska Pipeline System (TAPS) was constructed between 1974 and 1977 in response to the 1973 oil crisis. It conveys oil from Prudhoe Bay in northern Alaska to the port of Valdez in the southeast, a distance of over 800 miles (1,300 km). Building the pipeline system meant dealing with a multiplicity of complex design and management decisions that involved engineering, environmental, political, legal, security, financial, and other issues. A decision was made to run most of the pipeline above ground, supported by permafrost, which engendered an innovative and creative set of solutions. An interesting major concern was to find a way not to interfere with the annual caribou migration. Security was (and is) a big issue. Some unanticipated risks also arose, some with unintended consequences. This paper examines the responses to myriad challenges, examining it from a systems engineering and systems thinking viewpoint. Questions for discussion are suggested so that this can be used as a case study in a course on systems engineering or systems thinking.

1 Introduction

The United States geological survey has estimated that areas north of the arctic circle have up to 90 billion barrels of oil available in 25 areas (including offshore), but finding practical means of production and transportation of crude oil in these harsh conditions present difficult challenges. (A barrel, abbreviated bbl, contains 42 U.S. gallons, or 159 L.)

In June of 1968, a joint venture of ARCO and the Humble Oil and Refining Company announced the discovery of *recoverable reserves*—oil that is technically and financially feasible to extract—of 5–10 billion barrels in Prudhoe Bay in

R.S. Swarz (✉)
Worcester Polytechnic Institute, 100 Institute Road, Worcester, MA 01609, USA
e-mail: rswarz@wpi.edu

G. Fanmuy et al. (eds.), *Complex Systems Design & Management*,
DOI 10.1007/978-3-319-49103-5_2

northern Alaska. The climate there is severe: the average daily mean temperature is –21 °F (–29 °C) in February (the coldest month) and 47 °F (8 °C) in July (the warmest).

This area of Alaska, known as The North Slope, has a tundra climate. Prudhoe Bay is home to thousands of migratory birds, caribou, and other wildlife. It is also the largest oil field in the United States. The Trans-Alaska Pipeline System connects this field with a year-round navigable marine terminal in the south of Alaska via a 48″ (122 cm) diameter pipe, which runs through over 800 miles (1,300 km) of Alaskan wilderness. About half of the pipeline is above ground, pictured below.

Building the pipeline system meant dealing with a multiplicity of complex design and management decisions that involved engineering, environmental, political, legal, security, financial, and other issues. A decision was made to run most of the pipeline above ground, supported by permafrost, which engendered an innovative and creative set of solutions. An interesting major concern was to find a way not to interfere with the annual caribou migration. Security was (and is) a big issue. Some unanticipated risks and opportunities arose, some with unintended consequences.

The pipeline's statistics are staggering: It can hold over 9 million barrels of oil and is currently pumping approximately 200 million bbl/day. Historic throughputs have exceeded 750 million bbl/day! Since its inception, it has pumped more than 17 *billion* barrels of crude oil and has supported over 100,000 jobs in Alaska.

The motivation for building the pipeline has primarily political roots: During October 1973, there was a war between a coalition of Arab gulf states and Israel, which began with an Arab sneak attack on Israeli positions. Israel had anticipated and was well-prepared for such an eventuality, so the war lasted less than 3 weeks; however, there were far-ranging implications to the cost of the world's oil.

In an initial protest to the United States' support of Israel in this war, the Arab members of the Organization of Petroleum Exporting Countries (OPEC) reduced their oil production rate by 5 % almost immediately. Then, when President Nixon ordered additional military support to Israel, Saudi Arabia led OPEC to declare a complete embargo of oil going to the United States, Canada, Japan, the Netherlands, and the United Kingdom. The primary result of the embargo to worldwide oil prices was swift and dramatic, leading to a quadrupling of the price of oil and directly inspiring the interest in building the pipeline. The engineering, environmental, and other challenges that arose were daunting.

The estimated cost of the pipeline when it was first proposed in 1969 was $900 million. Within a year, that estimate had risen by 122 % to $2 billion. By 1973, that estimate rose again to a range of $3–4 billion and then to $7.7 billion by 1976. The final cost was estimated to be $8 billion. Construction of the Valdez Marine Terminal cost an additional $1.4 billion.

2 Challenges

This was a very complex project involving many challenges, which make an ideal case study for systems engineering and systems thinking. Costs were consistently underestimated. Challenges from native people and conservationists—the final Environmental Impact Statement of 1972 ran to 6,500 pages in 9 volumes—were numerous and strong. The engineering challenges of supporting the pipeline above ground on permafrost were unprecedented.

In the end, all arguments against the pipeline were ultimately rejected and the pipeline was built. Following is a more complete description of some of the more significant challenges.

2.1 Technical

From a systems engineering perspective, there were many architectural and design alternatives which needed to be analyzed to assess the very difficult technical challenges of transporting oil from Prudhoe Bay to Valdez. Oil emerges from the ground at temperatures as high as 160 °F (71 °C). Even though it cools a bit over its 800-mile journey, fluid friction tends to keep the temperature up. Oil pipelines in less severe environments have no problem in sending the oil at elevated temperatures, but the most significant problem for this Alaskan oil is that the subsoil on the route consists mainly of *permafrost*, which is defined as rock or soil material that has remained below 32 °F (0 °C) continuously for two or more years.

Running the pipe along permafrost, be it buried or above ground, presents difficult engineering challenges, because there is no solid ground on which to support it and the pipes are hot. In more temperate climates, buried pipe would not soften or melt the surrounding ground. Above-ground support structures would normally rest on existing rock or concrete pads, but no such thing is possible in Alaska because of the permafrost. The supporting structure could rest on permafrost; however, it must not be allowed to get warm, lest it melt the permafrost. The solution that was devised was to have vertical support members (VSM) made of 18″ steel pipe placed every 50–70 feet along the pipeline. Each pair of pipes had a cross member (pictured to the right) with a Teflon base that allows lateral movement in the case of expansion and contraction and seismic activity.

"Thermal" VSMs are used on most of the above-ground sections. These have pairs of 2″ pipes running from the base below ground to aluminum heat radiators at the top. The pipes contain anhydrous ammonia refrigerant which carries heat away from the permafrost and recycles itself without requiring any sort of control system.

All types of oil lines need to be cleaned constantly and checked for corrosion. This is accomplished in several ways. First, at the head end, prior to oil entering the main pipeline, water and gas is removed from the oil. Second, corrosion-inhibiting chemicals are added to the oil before it goes into the main pipeline. In the main pipeline itself, devices known as "pigs," shown to the right, are inserted into the pipeline and are pushed through it by the flow of oil Some pigs just scrape and clean the walls of the pipe. Other pigs—so-called "smart pigs"—can test things like the extent of corrosion and the thickness of the pipe wall.

2.2 *Political*

Political and environmental concerns began campaigns that successfully halted pipeline construction from 1970 to 1973.

Recall that Alaska was purchased by the United States from Russia in 1867 for $7.2 M, in a deal brokered by Secretary of State William Seward, which was at the time ridiculed as "Seward's Folly." In 1902, prior to statehood, the U.S. Department of Agriculture set aside 16 million acres (64,750 km^2) as the Tongass National Forest. An Alaskan native group, the Tlingits, believed that the land belonged to them and attempted to sue for its return. In 1959, Alaska became the 49th state under President Dwight Eisenhower.

A cash settlement of \$7 M was offered and rejected. A group called the Alaska Federation of Natives suggested that a more appropriate settlement should include \$500 M and 40 million acres. Under President Richard Nixon, this group agreed to abandon its land claims in favor of a settlement of nearly \$1B and 148.5 million acres (601 thousand km^2).

3 Alternatives Considered

As now built and functioning, the pipeline consists of 800.3 miles (1,288 km) of stainless steel pipe, 48″ (122 cm) in diameter. 420 miles (676 km) of the pipe are elevated on 78,000 supports that descend into the permafrost and have a unique system to support the pipe above ground while the supports are resting on permafrost.

Befitting of a systems engineering approach, before a solution was chosen, multiple alternatives were suggested and considered, as listed below.

3.1 The Boeing RC-1

The Boeing Corporation proposed the development of a mammoth transport aircraft. It was to have a wingspan of almost 478′ (146 m). In contrast, the largest cargo aircraft in service today is the Antonov An-124, with a wingspan of about half that. The RC-1, as it was called, was to be powered by 12 Pratt and Whitney JT9D jet engines. The RC-1 would have been about twice the size and weight of the An-124, but would have carried about five times the payload.

A unique part of the design was the runway system that was conceptualized, which consisted of three parallel, simultaneously utilized runways. The outside runways were used for landing only and the center strip was used as a taxiway. Aircraft unloaded at the end of the runway. The lightened plane could easily take off on the downwind.

3.2 Submarines

Another fascinating design was proposed by the General Dynamics company and consisted of a proposed fleet of submarines that would navigate under the polar ice caps. There would be a total of 17 boats, each costing \$700 M for the conventionally powered modes. Shore facilities would add another \$2–3B.

Nuclear-powered versions were also considered. They would cost an additional $25 M each, but because the nuclear subs are faster, only 14 boats would be needed.

It was suggested that that using submarines would have provided a significant additional political advantage, too, in that the subs could travel undetected to a variety of ports, depending on any current geopolitical situation. In the end though, the fuel consumed and other maintenance and operational expenses made this solution impractical.

3.3 Extension of the Alaska Railroad

The Alaska Railroad has its southernmost terminus at Seward—about 125 miles (201 km) south of Anchorage, and going northward from there connects the state's main population centers of Anchorage and Fairbanks, where it terminates, a distance of approximately 470 miles (756 km). In order to serve the pipeline, track would have had to be built on the northern end from Fairbanks to Deadhorse (at Prudhoe Bay) and on the southern end from Seward to the oil tanker terminal at Valdez on Prince William Sound. Despite being the snowiest city in the United States, Valdez has a long history as a commercial fishing port, navigable year-round.

3.4 Ice-Breaking Tankers

In 1969, Humble sent a specially modified oil tanker, the *Manhattan*, to test the theory that ice-breaking tankers could be used to transport the oil through the Northwest Passage from the Atlantic Ocean to the Beaufort Sea. Although the westward journey was completed, multiple cargo compartments flooded, and the course had to be changed during mid-journey due to extreme weather conditions. Canadian Coast Guard ice-breaking cutters escorted the Manhattan on its return trip.

The *Manhattan* was able to make a second successful trip in the summer of 1970, but the experiment was, nevertheless, viewed as a failure. There was simply too much risk of human casualties and oil spills.

4 Risks and Unanticipated Consequences

4.1 Security Risks

Obviously, the security of the pipeline is a clear and increasing concern. The *Anchorage Daily News* reported in 2006 that an organization affiliated with Al

Qaeda put postings on their web site that encouraged attacking the pipeline with bullets or explosives.

In one of the better-known incidents, an Alaska resident shot one hole into the pipeline in 2001. This attack took place on a section that had particularly high pressure, and the result was a plume of oil that rose 75 feet into the air! Almost 300,000 gallons of oil was spilled before the system was shut off. A crew efficiently repaired the break and restored normal flow within three days. The cleanup of the tundra, of course, took much longer. The photograph to the right is a patch to a hole in the pipeline caused by a bullet.

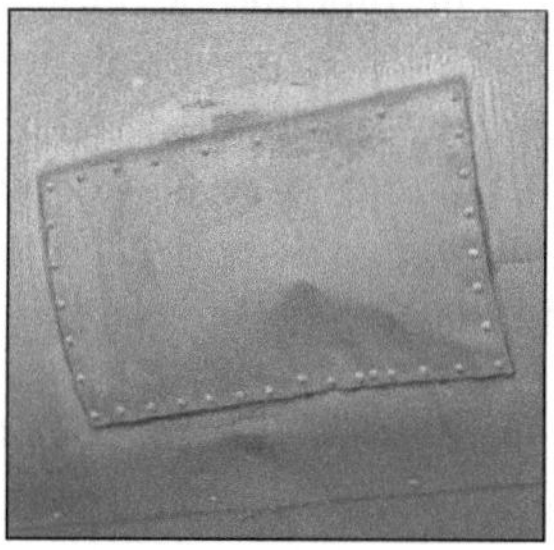

The current security system consists of fences, armed guards, and access controls at the pump stations and other vulnerable facilities; periodic aerial and ground patrols of the pipeline; intrusion detection systems at some facilities; and an emergency communications system. Alyeska, the pipeline operator, has plans in place to expeditiously involve federal and state law enforcement agencies for assistance if necessary. Security and oil spill assessment exercises have been conducted with satisfactory results.

In the event of deliberate attack or unintentional leaks, there are three redundant leak detection systems:

- A system which compares the amount of oil entering the pipeline with that exiting it
- A system which compares calculated flow with reported flow
- A system of flow and pressure sensors than can detect and localize anomalies

4.2 Unanticipated Consequences

One of the biggest concerns from the public was interference with caribou migration. There are two large caribou herds, each now numbering in the tens of thousands, in the Alaska National Wildlife reserved, the “Porcupine” and the

Central Arctic. Each year, in early March, the herds gradually migrate northwards towards the oil fields. Not impeding their migration is one of the reasons for the above-ground pipeline. Opponents feared that the pipeline would negatively impact their migration and threaten their very existence. Surprisingly, the herds have flourished! In 1977, the Central Arctic heard was estimated to be about 6,000—it is now estimated to be over 27,000. It is suspected that the heat generated by the pipeline makes a better environment for calf-bearing.

On the human side, there were some significant negative unanticipated consequences in Fairbanks, which became the center for hiring pipeline construction staff and to warehouse equipment. The pipeline developers were paying top dollar for construction staff, well over twice the existing salaries. This was a strong incentive for hordes of prospective employees to flock to Fairbanks. The population of Fairbanks doubled between 1970 and 1975 and continued to grow thereafter. But the additional population and the wage disparity led to disproportionate increases in the cost of clothing, food, and housing and to a large increase in all types of crime, including violent.

Native Americans were also adversely impacted. The pipeline development company was required by law to hire at least 3,000 Native Americans. After the pipeline's construction, many of these employees returned to their villages, after having received a sometimes ten-fold increase in their salaries. This change in lifestyle and cultural integration was difficult for many to resolve and led them to abuse alcohol and drugs and to abandon their native culture. Many subsequently left their native villages, which suffered deeply from the decreased population.

5 Conclusions

Despite the difficulties and dire projections, the pipeline has been enormously successful.

Consider that:

- Concerns about preventing migration seemed to be wrong. In fact, the Central Arctic herd which numbered about 6,000 in 1977 grew to over 27,000 by 2006. The long-term impacts are unknown.
- Fears about the effect of earthquakes were alleviated when a 7.9-magnitude quake struck on the Denali fault, described as one of the largest earthquakes in American history. Although there was some minor damage to the pipeline, it did not rupture.
- Although there have been a number of pipeline spills over the years, none have been devastating. (The Exxon Valdez spill is not considered to be pipeline-related.)

- Oil production has been as robust as anticipated. The chart below shows the annual throughput from 1977 to 2015.

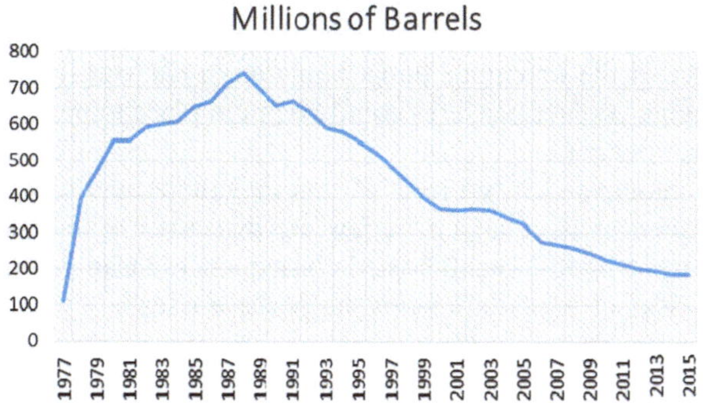

6 Questions for Discussion

Question 1: Advances in systems engineering are often the result of either technology "push" or competitive advances in technology. At other times, they are clear outgrowths of non-technical factors, such as political, economic, or environmental ones, such as the development of electric automobiles. Can you suggest some other examples of advances in system design or systems engineering whose development was inspired by non-technical factors?

Question 2: Perform a "back of the envelope" trade study of the four proposed alternatives, any other obvious alternatives you can think of, plus the solution chosen. What would have been the most important criteria? How would they be prioritized? What information about each alternative would be required?

Question 3: The motivation for building the Trans-Alaska pipeline and the responses to the various challenges—political, environmental, and technical—were numerous. What do you think would have happened if there hadn't been a war, if the conservationists hadn't raised concerns about caribou migration and other effects? Would the pipeline have been built at all? Would the technical challenges have been as great? Can you think of other scenarios and solutions?

Question 4: The Keystone pipeline consists of several operational stages and a proposed expansion segment, Keystone XL. The existing segments total 2,151 miles (3,461 km) and carry Canadian crude oil the U.S. Midwest and Oklahoma. All of it is buried at least 4′ (122 cm). If constructed, it would consist of several additional segments. If the XL project is ever completed, it would carry American crude oil from Baker, Montana to Cushing, Oklahoma. This is a highly-charged political situation. President Obama is against Keystone XL, largely because of fears of effecting climate change. What lessons, if any, from the Trans-Alaska Pipeline System can be applied to Keystone XL?

Question 5: The security of the pipeline currently depends upon planning for incidents, periodic inspections, dedicated communications, and traditional "guns, gates, and guards." In light of the changing threat situation, what vulnerabilities may now be exposed and what countermeasures could be put in place to protect the oil and the environment from attack?

Question 6: As can be anticipated in projects of this magnitude and diversity, many unanticipated consequences have arisen. A few are discussed in this paper. Can you envision any other potentially unanticipated consequences? Can you think of any mitigations for the risks or exploitations of the positive ones?

Bibliography

1. Alyeska Pipeline Service Co.: Pipeline Operations: Throughput. http://www.alyeska-pipe.com/TAPS/PipelineOperations/Throughput. Accessed 23 April 2016
2. An Analysis of the Economic and Security Aspects of the Trans-Alakka Pipeline: U.S. Department of the Interior, Assistant Secretary—Program Policy, Office of Economic Analysis, Washington DC, December 1971
3. Anderson, R.W.: Alaska Pipeline Doomsayings [sic] Revisited, 1/17/2006. http://www.mrc.org/news/alaska-pipeline-doomsayings-revisited. Accessed 27 April 2016
4. Arctic Power ANWR Information Brief: Do the Caribou Really Care? http://www.anwr.org/features/pdfs/caribou-facts.pdf. Accessed 22 April 2016)
5. Banet, A.C.: Oil and Gas Development on Alaska's North Slope: Past Results and Future Prospects. Bureau of Land Management (1991)
6. Canfield, M.: Statement before the U.S. Senate Committee on Energy and Natural Resources, September 26, 1977
7. Chance, N.: The Arctic National Wildlife Refuge: A Special Report. http://arcticcircle.uconn.edu/ANWR. Accessed 15 April 2016
8. Coates, A.P.: The Trans-Alaskan Pipeline Controversy: Technology, Conservation, and the Frontier. University of Alaska Press, Fairbanks (1993)
9. Jacobs, D.: The caribou question, vol. 19, No. 2. Property and Environment Research Center (2001)
10. Lovins, A.B., Lovins, L.H.: Brittle Power: Energy Strategy for National Security. Brick House Publishing Company (1982)
11. McKibben, B.: One guy, one rifle, and an oil pipeline. Los Angeles Times (2001)
12. Moore, S., Griffith, J.: Lessons for the keystone XL pipeline debate. The Heritage Foundation Backgrounder #2977 on Energy and Environment (2014)
13. Pipeline Facts: Alyeska Pipeline Service Company. http://www.alyeska-pipe.com/TAPS/PipelineFacts. Accessed 19 April 2016
14. Reichart, D.: Positive and Negative Externalities. https://mba651fall2007.wikispaces.com/Positive+and+Negative+externalities. Accessed 21 April 2016
15. Robert Douglas Mead: Journeys Down the Line. Doubleday & Co., New York (1978)
16. Roscow, P.J.: 800 Miles to Valdez: The Building of the Alaska Pipeline. Prentice-Hall, Englewood Cliffs (1977)
17. Strohmeyer, J.: Extreme Conditions: Big Oil and the Transformation of Alaska. Todd Communications (1997)
18. The ice above, the giant below. Popular Science, p. 182 (1982)

19. United States General Accounting Office: Report GEO/RCED-92-58BR, Trans-Alaska Pipeline: Insuring the Pipeline's Security (1991)
20. Vogely, W.A.: An Analysis of the Economic and Security Aspects of the Trans-Alaska Pipeline. U.S. Department of the Interior, Office of Economic Analysis (1971)
21. Wikipedia Contributors: 1973 oil crisis. Wikipedia, The Free Encyclopedia. https://en.wikipedia.org/w/index.php?title=1973_oil_crisis&oldid=714746302. Accessed 19 April 2016
22. Wikipedia Contributors: Boeing RC-1. Wikipedia, The Free Encyclopedia. https://en.wikipedia.org/w/index.php?title=Boeing_RC-1&oldid=665633524. Accessed 19 April 2016
23. Wikipedia Contributors: Keystone pipeline. Wikipedia, The Free Encyclopedia. https://en.wikipedia.org/w/index.php?title=Keystone_Pipeline&oldid=714186652. Accessed 25 April 2016
24. Wikipedia Contributors: Oil reserves. Wikipedia, The Free Encyclopedia. https://en.wikipedia.org/w/index.php?title=Oil_reserves&oldid=713347191. Accessed 19 April 2016
25. Wikipedia Contributors: Prudhoe bay oil spill. Wikipedia, The Free Encyclopedia. https://en.wikipedia.org/w/index.php?title=Prudhoe_Bay_oil_spill&oldid=691263249. Accessed 19 April 2016
26. Wikipedia Contributors: Trans-Alaska pipeline system. Wikipedia, The Free Encyclopedia. https://en.wikipedia.org/w/index.php?title=Trans-Alaska_Pipeline_System&oldid=710757698. Accessed 19 April 2016
27. Wikipedia Contributors: Yom Kippur war. Wikipedia, The Free Encyclopedia. https://en.wikipedia.org/w/index.php?title=Yom_Kippur_War&oldid=715794890. Accessed 19 April 2016

MBSE, PLM, MIP and Robust Optimization for System of Systems Management, Application to SCCOA French Air Defense Program

Thomas Peugeot, Nicolas Dupin, Marie-Joëlle Sembely and Catherine Dubecq

Abstract To examine the Project Management aspects of the French Air Defense Program SCCOA, a Model-Based System Engineering approach using the NATO Architecture Framework (NAF) is appropriate to ensure the System of Systems consistency. Two limitations of the NAF are addressed: incorporating temporality and incorporating decision support tools. The first issue is resolved by coupling NAF with an Access calendar database. The second is solved using Prolog, a Constraint Programming tool, and Cplex, a Mathematical Programming tool. The resulting tool stack allows to schedule deployment integrating Robust Optimization techniques.

1 Introduction

The French Defense Procurement Agency of the French Ministry of Defense (DGA) needs to manage complex systems both from a human and organizational standpoint. There is a clear willingness within the DGA to deploy System Engineering (SE) and Model-Based System Engineering (MBSE, see [1]) on a large scale in order to rationalize their decision making. Operations Research (OR) can provide decision support tools in this context in interface with Product Life-cycle Management (PLM). This article focuses on the deployment and the experience feedback of such methodology for the French Air Defense program SCCOA [2].

T. Peugeot (✉) · C. Dubecq
MOSS S.A.S., 86 rue Henri Farman, 92130 Issy Les Moulineaux, France
e-mail: thomas.peugeot@moss.fr

C. Dubecq
e-mail: catherine.dubecq@moss.fr

N. Dupin (✉)
Direction Générale de l'Armement (DGA),
60 Boulevard du Général Martial Valin, 75015 Paris, France
e-mail: nicolas.dupin.2006@polytechnique.org

M.-J. Sembely
Airbus Defence and Space for MOSS SAS, Les Mureaux Cedex, France
e-mail: marie-joelle.sembely.ext@moss.fr

G. Fanmuy et al. (eds.), *Complex Systems Design & Management*,
DOI 10.1007/978-3-319-49103-5_3

The System of Systems SCCOA The French Air Defense program SCCOA is composed of hundreds of deployed systems from radars to command/control centers with telecommunication artifacts. SCCOA is an example of a system-of-systems (SoS) as described by [3, 4]: it is an assemblage of systems that can be acquired and/or used independently, for which the designer tries to maximize the performance of the global value chain at a given time and for foreseeable assemblages. SCCOA's Project Management (PM) is decomposed into several management decision levels for detection, telecommunication and fixed centers, deployable systems ... SCCOA is furthermore interfaced with other military and civilian programs, which makes SCCOA's PM particularly rich in interfaces. To face such complexity, SCCOA's PM is decomposed incrementally to renew SCCOA's fleet, the actual increment is SCCOA 4.2 whereas the next in preparation is SCCOA 5.

Model-Based System Engineering and SoS It is well known [5] that disseminating documentation which contains inconsistencies is a failure factor in large systems. SCCOA has adopted MBSE to avoid this risk. The benefits of MBSE are substantial because designed models are integrated into a shared repository. When designing new systems, MBSE ensures the consistency from different viewpoints: capability, operational, system and service. The contents of such repositories are used to generate consistent system documentation, we refer to [6, 7].

The NATO Architecture framework (NAF) is one of the current MBSE enabler that provides common language and structure for defense projects (similar to DoDAF and MODAF). However, one drawback of NAF is its complexity (280 concepts, 48 views). Developing a domain specific ontology on top of the NAF metamodel [8, 9] allows to share informations, concepts and structures efficiently.

When applied to a SoS, a MBSE solution needs to take into account deployments and retirements.

PLM tools are appropriate to manage spatial and temporal configuration data. The key question is therefore how to create an effective interface between MBSE and PLM tools. While such an interface has already been mentioned in [10], it has not to our knowledge been studied at length for SoS.

State of the art in Optimization in System Engineering PLM issues naturally give rise to optimization problems, as mentioned in [11, 12]. Multi-objective optimization furnishes best compromise solutions in a Pareto front and is an appropriate framework to deal with concurrent objectives (e.g. cost and robustness), we refer to [13, 14]. Evolutionary algorithms are commonly used for a resolution framework to solve such problems, we refer to [11, 15, 16]. As a drawback, evolutionary algorithms require very specific implementation and parametrization. It requires some specific expertise and significant efforts to be efficient, even with the unified implementation provided in [17]. Recent progress in OR led to efficient model&run solvers where the implementation relies on a black box solver to focus on a modeling work. For instance, LocalSolver [18] has model&run facilities with a black box resolution based on an aggressive local search algorithm. Constraint Programming (CP) is another model&run paradigm for constrained optimization problems. CP has already been used successfully to tackle SE issues, we refer to [19, 20]. Mixed Integer Programming (MIP, we refer to [21]) is an exact optimization framework allowing a model&run implementation and optimality guarantees. MIP was also used to tackle SE issues in [12, 22]. Tractable resolution sizes are more limited using MIP than meta-heuristics. However recent advances allow to use MIP for large industrial problems, both to find solutions heuristically and to compute optimality gaps we refer to [23–25]. Recent advances in OR also allows to consider uncertainty in the input data of MIP problems, we refer to [26] for Robust Optimization.

Paper outline This paper is organized as follows. Section 2 introduces the specific problematic of system retirements giving an overview of the tool stack elaborated in response to this. Sections 3–6 introduce the specific modeling details of the component of the tool stack: the MBSE architecture, the pivot ontology, the MIP formulation of the problem, and the robust optimization extensions. Section 7 presents implementation issues and experience feedbacks. Section 8 summarizes our contributions and opened perspectives.

2 Industrial Problematic and Solution Outline

Problem presentation The tool stack presented in this paper was created in response to the migration of network artifacts towards new technologies compatible with Internet Protocol, illustrated in Fig. 1. Scheduling such a transition calls for a clear understanding of topology (what type of data flows between systems) and configurations, (which versions can support the new communication technology). Because these systems are procured and managed by different agencies/teams, the project scheduling involves working with loosely-coordinated decision makers. Currently the constraints linked to the deployment schedule are perceived as difficult because

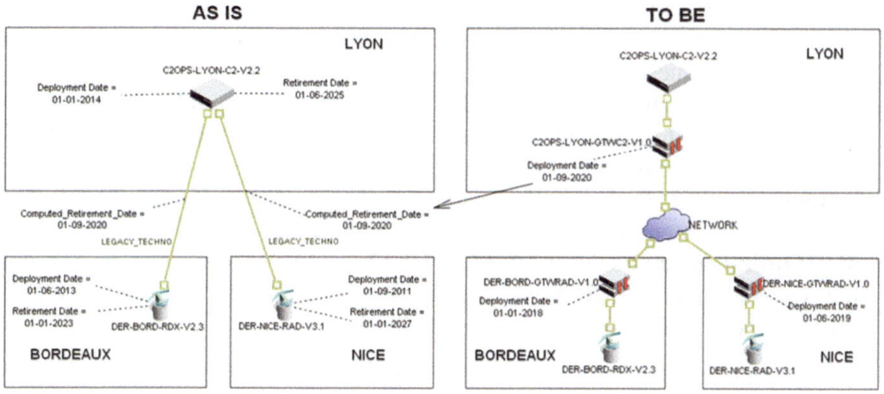

Fig. 1 As is/to be, illustration of the migration of telecommunication artifacts. These pictures are not obtained with the real data for the sake of confidentiality, but are representative of the problematic and correspond exactly to the view provided by the tool

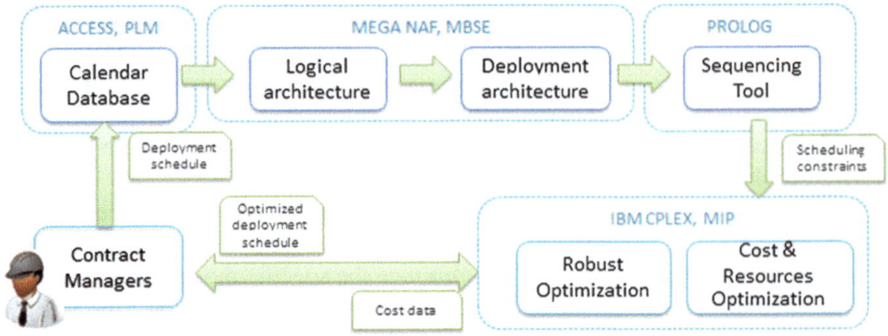

Fig. 2 Tool stack supporting the decision process

no tool is available to manage technical, temporal and financial constraints. Furthermore uncertainties such as delays in project milestones have an adverse effect on scheduling.

The tool stack In response to this difficult problematic, a tool stack from MBSE and PLM to OR illustrated in Fig. 2 has been addressed, integrating state of the art technologies. All the tools are based on COTS and communicate between themselves. MBSE helps to capture elements from different agencies/teams and ensures consistent data, PLM addresses the temporality and OR furnishes tools to rationalize complex decisions. The following tools were developed:

- **Calendar Database**: this is a contract driven database. Each system deployment or retirement operation is captured and updated during monthly reviews with contract managers. A level of confidence is associated to each date in the calendar database. This lightweight PLM tool is implemented with MS ACCESS.
- **Logical Architecture**: it is implemented with a MEGA NAF model described in Sect. 3. This model is common for SCCOA's SoS management and is a basis for all opportunity questions.
- **Network Architecture**: a distinct MEGA NAF model to extend the modeling for specific needs of the problematic of links dismantling. Network and Logical Architecture are synchronized with correspondence rules.
- **Sequencing Tool**: Prolog [27] provides sequencing constraints and the earliest possible deployment dates. It is implemented in Prolog.
- **MIP Optimization**: Based on IBM CPLEX [28], a MIP model optimizes scheduling with additional constraints including a financial model. It allows to take into account uncertainties related to project milestones.

Having this global view of the tool stack in response to the industrial problematic, the next sections focus on the specific modeling aspects of MBSE, PLM, OR frameworks and the integration and interface questions.

3 MBSE for the System of Systems SCCOA Architecture

Our MBSE modeling followed the practices previously outlined. An ontology was developed to capture SCCOA's capability, operational and system views. We note an originality compared to usual deployments of MBSE approaches: our MBSE model concerns at this stage the modeling of the existing systems and configurations in the increment SCCOA 4.2 as a preliminary work for further development to design the architecture for SCCOA 5. Figure 3 illustrates the SoS ontology on top of NAF where the following concepts are modeled:

- **Capabilities**: Two levels of capability are defined. The first level organizes capabilities. The second level provides specific elements that can be evaluated.
- **Functional Chains**: Capabilities are linked to system process embodied in Functional Chains with NSV-4 views. A Functional Chain is an assembly of system performing functions for a specific mission.
- **Operational Activities**: NOV-2 views provide a consistent operational vocabulary across the SCCOA.
- **Operational Centers**: A NOV-5 map of operational centers is provided, operational activities are mapped to those centers.
- **Systems**: Systems are linked to Functional Chain via their System Function and are deployed on Operational Centers.
- **Deployed Centers**: The deployment of systems in centers is also modeled. It is not mandatory for a SoS model, it was a need for our problematics.

Because of the vast variety of operations accomplished using SCCOA, no single expert is able to validate the whole model. Since the priority is to capture validated models, the SoS is broken down into Functional Chains. This breakdown matches the areas of expertise of the subject matter experts (SME) who can validate the model. Each of those SMEs plays his own key role within those Functional Chains. Most of them are not familiar with modeling practices. Formal reviews of the Functional Chains allow engineers and operational staff to share their insight.

The choice to validate functional chains is an illustration of a *middle-out* approach. A middle-out approach combines indeed top-down and bottom-up approaches: it provides a coordination of stakeholders with common concepts and languages (like in top-down approaches), with the possibility for SMEs to have an influence on high level choices (like in bottom-up approaches).

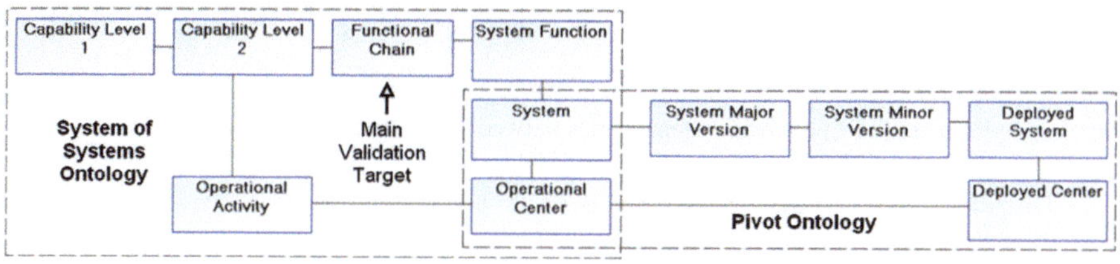

Fig. 3 SoS and Pivot ontology on top of the NAF metamodel

Operational views (NOV-x) or System Architecture views (NSV-1) cannot be submitted for review because there are deemed too abstract by the SMEs. Functional Chains views (NSV-4) are understandable with little preparation because they combine systems and process in one view. Therefore, Functional Chains form the main part of the validation process.

4 Pivot Ontology for Communication Between Tools

Pivot ontology The tool stack shares a common ontology to enable information exchange as described in Fig. 3. *Deployed Systems* are described in the Calendar Database. For each operation, a fine grain configuration (the *System Minor Version*) is deployed on a given *Deployed Center*. Because the Calendar Database can describe too many minor versions of a system, the concept of *System Major Version* captures major changes in the system (usually 2–3 major versions per system). Those changes between System Major Version are modeled in the Logical/Network Architecture (for instance when a system supports a new protocol).

Enforcement of correspondence rules Our experience is that the correspondence rules between data in the tool stack can be established without too much effort using the "model and run" facilities of SWI-Prolog tool [27]. As stated in ISO/IEC/IEEE 42010 "an architecture description (AD) element is any construct in an architecture description. A correspondence defines a relation between AD elements. Correspondences are used to express architecture relations of interest within an architecture description". In our case, correspondences are specifically written for each pair of tools. Figure 4 describes the enforcement of correspondence rules between contiguous tools in the tool stack with Prolog. A Prolog environment is a set of Facts (Knowledge Database) and Rules (the Rules database). Knowledge databases are automatic exports of the tool.

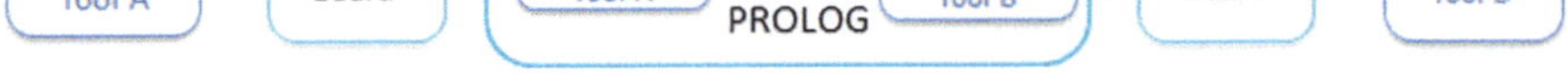

Fig. 4 Implementation of correspondence rules between Architecture Descriptions

5 MIP Optimization to Schedule SCCOA's Deployments

This section focuses on the MIP modeling to optimize the financial cost to dismantle old links, with resource constraint limiting deployment.

Notations We denote with $s \in \mathscr{S}, c \in \mathscr{C}, l \in \mathscr{L}$ the indexes and sets to designate respectively SCCOA sites, command and/or control centers and links to dismantle. SCCOA links are partitioned with $\mathscr{L} = \mathscr{L}_1 \cup \mathscr{L}_2$, $\mathscr{L}_1$ being new liaisons to deploy whereas $\mathscr{L}_2$ are the old liaisons to dismantle.

$d \in \mathscr{D}$ denotes new SCCOA systems to deploy, with deployment date in $[d^b, d^f]$. Deployment of new systems requires resource usages that limit the simultaneous deployments. Δ_r denotes the usage duration for new systems $\mathscr{D}_r \subset \mathscr{D}$ requiring resource r, $n_{r,t}$ being the maximal number of resources r available at time step t.

Financial costs can be associated to scheduling decisions. We denote with $k \in \mathscr{K}$ the old obsolete technologies. $\mathscr{K}_l \subset \mathscr{K}$ designates the subset of technologies that requires the link to dismantle $l \in \mathscr{L}_2$. When a technology k is still used, the cost/time unit to maintain it is C^k. To have the earliest dismantling of sites, we associate cost/time unit for the not dismantled sites s with C^s.

Financial costs versus minimizing completion time Our problematic has similarities with the academic OR problem RCPSP [29]. RCPSP schedules jobs with resource and precedence constraints minimizing the completion time to realize all the jobs. Cost optimization is not equivalent to the earliest scheduling similarly to the RCPSP: Fig. 5 illustrates that optimal scheduling regarding financial costs can be non optimal regarding the minimization of completion times.

MIP modeling The problem can be formulated in MIP as following. Continuous variables T^s, T^c, T^l are introduced to indicate dismantling dates for sites s, centers c, links l. Binary variables $x_{d,t} \in \{0, 1\}$ are defined for all new systems to deploy and possible deployment date. To have an efficient MIP resolution as stated in [21, 30], $x_{d,t} \in \{0, 1\}$ are defined with $x_{d,t} = 1$ if system $d \in \mathscr{D}$ is deployed before t.

Equation (1) is the objective to minimize aggregating costs to maintain old technologies C^k and costs to have sites requiring old links C^s. $C^s = 0$ corresponds to financial optimization whereas $C^k = 0$ minimizes completion time. Constraints (2)–(6) are precedence constraints: (2) ensures that a site s can be dismantled once its centers c are dismantled, (3) ensures that each center is dismantled once all the related obsolete links are dismantled and the new links are operational. The dates related to

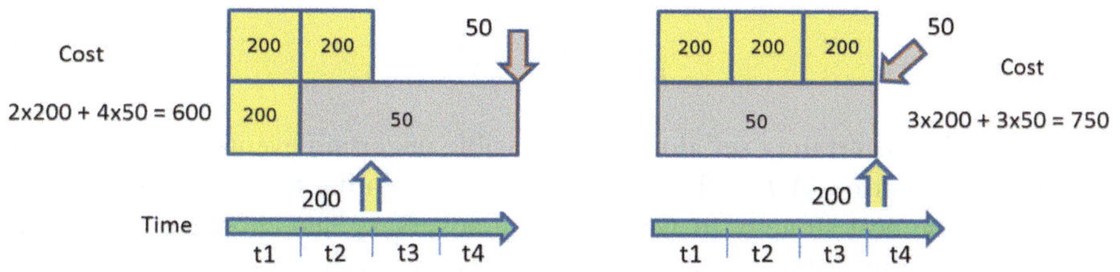

Fig. 5 Illustration that cost optimization is not minimizing the completion times

liaisons are coupled with related dismantling and installation of systems in (5) and (6). Equation (4) allows to compute the earliest dates when the different technologies are not used. its centers c are dismantled. Equation (7) are implied by the definition of variables x, (8) and (9) are the time windows to install new systems and (10) codes the resource constraints.

$$\begin{array}{rccr}
\min\limits_{x,T\geqslant 0} & \sum_s C^s(T^s - t_0) & + \sum_k C^k\ (T^k - t_0) & (1)\\
\forall c \in \mathscr{C}, & T^c & \leqslant T^{s_c} & (2)\\
\forall c \in \mathscr{C}, l \in \mathscr{L}_c, & T^l & \leqslant T^c & (3)\\
\forall l \in \mathscr{L}, k \in \mathscr{K}_l, & T^l & \leqslant T^k & (4)\\
\forall l \in \mathscr{L}_2, & T^l & \geqslant \min(t^{l_b}, t^{l_f}) & (5)\\
\forall l \in \mathscr{L}_1, d \in \mathscr{D}_l, t \in \mathscr{T}, & T^l & \geqslant t(x_{d,t} - x_{d,t-1}) & (6)\\
\forall d \in \mathscr{D}, t \in \mathscr{T}, & x_{d,t} & \leqslant x_{d,t+1} & (7)\\
\forall d \in \mathscr{D}, & x_{d,t_d^b-1} & \leqslant 0 & (8)\\
\forall d \in \mathscr{D}, & x_{d,t_d^f} & \geqslant 1 & (9)\\
\forall r \in \mathscr{R}, t \in \mathscr{T}, & \sum_{d\in\mathscr{D}_r}(x_{d,t} - x_{d,t-\Delta_r}) & \leqslant n_{r,t} & (10)
\end{array}$$

Resolution issues Without constraints (10), the problem contains only continuous variables and can be resolved polynomially using Prolog or a tool such as MS Project. With the resource constraints (10), the problem is NP-hard. This MIP model is powerful for decision making. Imposing end dates for T^s, T^c, T^l, MIP can quickly prove a calendar infeasibility. This model can also be used to analyze sensitivity to uncertain events, such as milestone delays.

6 Robust Optimization to Handle Data Uncertainty

Optimization under uncertainty was developed to search solutions that are resistant to some perturbations on initial data, jointly optimizing both costs and robustness. There is a wide variety of robust problems and case-by-case resolutions. We refer to [31] for a survey on robust scheduling. Several generic approaches can be implemented for the previous MIP.

General ideas Considering a deterministic MIP written as following:

$$\min_{\{x\in\mathbb{N}^m\times\mathbb{R}^p_+,\ Ax\geqslant b\}} cx \tag{11}$$

We note Ω the uncertainty set, domain of feasible uncertain events applying on the coefficient of matrices A, b, c. The robust problem is similar to a game theory problem, facing the best strategy in Ω of a fictive adversary that choose rationally the more penalizing uncertain event once the x decisions are played. It leads to the following min-max scheme:

$$\begin{aligned} \mathscr{P}^{rob} = & \min_{x \in \mathbb{N}^m \times \mathbb{R}_+^p} \max_{\omega \in \Omega} \quad c^T x \\ s.t: & \forall \omega \in \Omega, A(\omega)x \geqslant b(\omega) \end{aligned} \tag{12}$$

Linear Programming case The Linear Programming (LP) case, where there are no resource constraints (10) still has a polynomial resolution, defining an uncertainty set in a polyhedron, we refer to [32]. With $C^k = 0$, it is a special case of robust PERT scheduling studied in [33].

Cost uncertainty In the case where only cost coefficients of LPs or MIPs are uncertain, the approach of [26] applies generically.

Light robustness Light Robustness [34] is a heuristic decomposition of the min-max problem (12). The initial phase calculates the optimistic cost without uncertainty. Given the deterministic cost, it defines a threshold of "cost acceptable" solutions. The last phase computes the "most robust" solutions with the accepted over-cost. A Pareto Front can be computed to arbitrate best compromise solutions between cost and robustness, modifying the threshold of cost-acceptable solutions.

7 Implementation Issues, Experience Feedback

This section aims to point out implementation issues and experience feedback concerning the different tools and modeling frameworks.

MEGA NAF Using MEGA NAF was imposed by the contract for a deployment interoperability with DGA tools. Modeling SCCOA 4.2 with MBSE furnished already some returns on investment even if the main goal is to prepare SCCOA 5. The experience feedback is satisfactory with the automatic generation of chapters of analytical documents which ensures to have consistent documentation. Some vigilance points were raised thanks to the NAF SoS model to detect functional and/or temporal discontinuities. To detect vigilance points, some database interrogations and quality metrics were coded in Visual Basic.

The MBSE approach with MEGA NAF generates a website with selected views and informations. This consistent export from the NAF database had a clear success amongst stakeholders of SCCOA. The website is the most appropriate support to validate models, with clicking interactions and intuitive navigation allowing system architects to focus on the part of SCCOA they manage. The website gave satisfaction for its ability to capitalize knowledge amongst stakeholders, it was thus adopted for the training of newcomers in SCCOA.

Access, MS Project Access and MS Project are lightweight PLM tools, it was sufficient for the needs of the project. More complex PM tools could be interfaced in a tool stack mixing MBSE, PLM and OR tools. For a deployment of the tool stack in the DGA, Access and MS Project have the advantage to be widely used.

Prolog to implement pivot ontologies While Prolog is sometimes considered to be outdated, our experience was convincing. Having no initial knowledge of Prolog in the MBSE team, some moderate training is enough to take profit of powerful possibilities for a simple and clear implementation. Prolog is open source, and offers significant advantages. Correspondence rules are expressed with the concepts of the tools ontology (and not the NAF vocabulary). This enables a natural understanding of correspondence rules. Correspondence rules are concisely expressed in logical programming, which requires only a few lines of Prolog.

In retrospect, using Prolog to communicate between tool is useful: correspondence rules are nimble textual n-uplets, tools import change orders and export knowledge bases in the form of textual n-uplets. Since n-uplets are similar to Excel csv files, it is straightforward to develop import/export functions that adhere to knowledge bases. Therefore, Prolog points to a simplified OSLC [35] for integrating system engineering software.

Prolog is also used to code quickly some graph searches. Specific graph algorithms should be more efficient. Having satisfactory performances with quick developments encouraged us to use Prolog. For instance the generation of the first planning on the dismantling dates is a simple graph search, taking the maximum of the installation dates of new systems along the substitution path. Prolog is simple and generic, we recoded with Prolog the quality and continuity metrics formerly coded in Visual Basic for a better concision and code maintainability.

MIP optimization We used IBM Cplex for the MIP computations in a model&run implementation through the OPL interface, using OPL script for data preprocessing. MIP allows to deal with more complex models than Prolog taking into account resource constraints, financial costs and uncertainty with robust optimization techniques. Our numerical experiments show that the deterministic resolution to optimality is easy with Cplex. Other free and less efficient MIP solvers could be used efficiently. However, the simple implementation with OPL and OPL script were crucial for our development times.

Resource constraints (10) were not a limiting factor for our case: the MIP resolution proves that the first planning calculated by Prolog is still feasible. So the new decisions have few impacts in the dismantling planning which is mainly decided with the project milestones and the already planned decisions. This is not a disappointing result: MIP optimization proves the feasibility which interested the concerned manager. In this case, the robust resolution is equivalent to the worst case approach of [36], which can also be coded with Prolog. Light Robustness of [34] is also useful to provide a robust planning for the new deployments without over-cost.

One can discuss the choice of MIP optimization. MIP requires some specific expertise in modeling to be efficient, but offers optimality guarantees. For discrete optimization problems, we recommend for non experts to use more intuitive modeling frameworks such as LocalSolver [18] or Constraint Programming.

8 Conclusions and Perspectives

Conclusions Given the difficulties involved in managing the large System of Systems SCCOA, some NAF extensions were useful when deploying our MBSE approach. The first extension interfaces a lightweight PLM tool. The second extension models the specific characteristics of a SoS. The third extension is a bridge with a Mathematical Programming tool. These extensions make it possible to integrate a tool stack with state-of-the-art methods from operations research, to address deployment scheduling decisions. This tool stack gave satisfactory results to decision makers, and is currently used to renew deployment and retirement projections.

Perspectives To improve the tool stack, stochastic optimization seems promising. Simulation tools can also complete the tool stack once solutions are computed with optimization for sensitivity analyses. The tool stack designed for SCCOA's retirement and deployment planning uses generic methodologies and tools. A natural perspective would be to extend this methodology to other complex systems of systems. For the perspectives related to SCCOA, the preparation of the next increment SCCOA 5 will reuse the methodology and the NAF architecture model to design it. The need for OR tools will be in architecture optimization, robust optimization in network design is undoubtedly promising.

Acknowledgments This paper was written thanks to a team work, the authors are grateful to the stakeholders of System Engineering for SCCOA. Special thanks to Jean Reix and Jérôme Lemaire (SCCOA program directors), Lionel Roz (MOSS, director of the System Engineering service), Jean-François Dumas and Sara Sellos (SCCOA system architects). We also thank Lucas Hassan for his participation to code MIP models.

References

1. Estefan, J.A., et al.: Survey of model-based systems engineering (MBSE) methodologies. Incose MBSE Focus Group **25**(8) (2007)
2. SCCOA: http://www.defense.gouv.fr/dga/equipement/information-communication-espace/le-systeme-de-commandement-et-de-conduite-des-operations-aerospatiales-sccoa
3. Luzeaux, D.: SoS and Large-Scale Complex Systems Architecting. In: Complex Systems Design and Management, pp. 39–49. Springer International Publishing (2014)
4. Maier, M.W.: Architecting principles for systems-of-systems. In: INCOSE International Symposium, vol. 6, pp. 565–573. Wiley Online Library (1996)
5. Charette, R.N.: Why software fails [software failure]. IEEE Spectr. **42**(9), 42–49 (2005)
6. Luzeaux, D., Ruault, J.R., Wippler, J.L.: Complex Systems and Systems of Systems Engineering. Wiley (2013)
7. Simo, F.K., Lenne, D., Ernadote, D.: Mastering SoS complexity through a methodical tailoring of modeling: benefits and new issues. In: Systems Conference (SysCon), 2015 9th Annual IEEE International, pp. 516–520. IEEE (2015)
8. Ernadote, D.: An automated objective-driven approach to drive the usage of the naf framework. In: NATO Science and Technology Organization (STO) Symposium (2013)
9. Ernadote, D.: An ontology mindset for system engineering. In: 2015 IEEE International Symposium on Systems Engineering (ISSE), pp. 454–460. IEEE (2015)
10. Moones, E., et al.: Towards an Extended Interoperability Systemic Approach for Dynamic Manufacturing Networks: Role and Assessment of PLM Standards. In: CSD and M (2015)
11. Doufene, A., Chalé-Góngora, H.G., Krob, D.: Complex systems architecture framework: Extension to multi-objective optimization. In: CSD and M 2013, pp. 105–123. Springer (2013)
12. Helle, P., Masin, M., Greenberg, L.: Approximate reliability algebra for architecture optimization. In: Computer Safety, Reliability, and Security, pp. 279–290. Springer (2012)

13. Kim, I.Y., De Weck, O.: Adaptive weighted-sum method for bi-objective optimization: Pareto front generation. Struct. Multidisc. Optim. **29**(2), 149–158 (2005)
14. Smaling, R., Weck, O.D.: Assessing risks and opportunities of technology infusion in system design. Syst. Eng. **10**(1), 1–25 (2007)
15. Chen, M., Hammami, O.: A system engineering conception of multi-objective optimization for multi-physics system. In: Multiphysics Modelling and Simulation for Systems Design and Monitoring, pp. 299–306. Springer (2015)
16. Fleming, P.J., Purshouse, R.C., Lygoe, R.J.: Many-objective optimization: an engineering design perspective. In: Evolutionary Multi-criterion Optimization, pp. 14–32. Springer (2005)
17. Talbi, E.G.: Metaheuristics: From Design to Implementation, vol. 74. Wiley (2009)
18. Benoist, T., Estellon, B., Gardi, F., Megel, R., Nouioua, K.: Localsolver 1. x: a black-box local-search solver for 0-1 programming. 4OR **9**(3), 299–316 (2011)
19. Condat, H., Strobel, C., Hein, A.: Model-based automatic generation and selection of safe architectures. INCOSE (2012)
20. Sagaspe, L.: Allocation sûre dans les systmes aéronautiques: Modélisation, vérification et génération. Ph.D. thesis, Université Bordeaux 1 (2008)
21. Vielma, J.P.: Mixed integer linear programming formulation techniques. SIAM Rev. **57**(1), 3–57 (2015)
22. Marinelli, F., De Weck, O., Krob, D., Liberti, L., Mucherino, A.: A general framework for combined module-and scale-based product platform design. In: Second Internal Symposium on Engineering Systems MIT. Cambridge, Mass (2009)
23. Dupin, N.: Modélisation et résolution de grands problèmes stochastiques combinatoires: application à la gestion de production d'électricité. Ph.D. thesis, Lille 1 (2015)
24. Dupin, N., Talbi, E.G.: Dual matheuristic and new dual bounds for the EURO/ROADEF 2010 Challenge. IRIDIA Technical Report series (ISSN 1781-3794) (2016)
25. Dupin, N., Talbi, E.G.: Matheuristic for the discrete unit commitment problem with min-stop ramping constraints. IRIDIA Technical Report series (ISSN 1781-3794) (2016)
26. Bertsimas, D., Sim, M.: The price of robustness. Oper. Res. **52**(1), 35–53 (2004)
27. Wielemaker, J.: SWI-Prolog 2.7-Reference Manual (1996)
28. ILOG, I.: Cplex optimizer 12.6. 0 (2014)
29. Schulz, J.: Hybrid solving techniques for project scheduling problems. Ph.D. Thesis (2013)
30. Dupin, N.: Tighter MIP formulations for the discretized unit commitment problem with min-stop ramping constraints. To appear in EURO Journal of Computational Optimization
31. Herroelen, W., Leus, R.: Robust and reactive project scheduling: a review and classification of procedures. Int. J. Prod. Res. **42**(8), 1599–1620 (2004)
32. Remli, N.: Robustesse en programmation linéaire. Ph.D. thesis (2011)
33. Minoux, M.: Duality, Robustness, and 2-stage robust LP decision models. Application to Robust PERT Scheduling (2007)
34. Fischetti, M., et al.: Light robustness. Lect. Notes Comput. Sci. **5868**, 61–84 (2009)
35. OSLC: core specification version 2.0. Open Services for Lifecycle Collaboration (2010)
36. Soyster, A.: Convex programming with set-inclusive constraints and applications to inexact linear programming. Oper. Res. **21**, 1154–1157 (1973)

Disruptive Innovation in Complex Systems

The Ambition of Combining Systems Engineering and Design Thinking

Arnaud Durantin, Gauthier Fanmuy, Ségolène Miet and Valérie Pegon

Abstract For almost a year, the Design Studio and systems engineering teams at Dassault Systèmes have shared their respective practice: design thinking and complex systems engineering. This comparison gave us insights about several shifts: the people involved in project ecosystems, the call for more disruptive innovation, the growing capabilities of computers, the need to take into account the full complexity of humans and a few shared ambitions between both disciplines. After explaining this context, this paper reports on the comparison between the two practices, through a cross-referenced strength and weakness comparison, and other counterbalancing points. We also share early hypotheses, gleaned from our experiments, on how to combine the design thinking and systems engineering approaches in early stages of innovation, at the right time, despite cultural differences. To conclude, we look at what is needed to make complexity easier to grasp, how a combined approach also calls for a fresh look at project organisations and for a practice mixing art and technology.

A. Durantin · G. Fanmuy (✉)
Dassault Systèmes - Systems Engineering, Vélizy-Villacoublayc, France
e-mail: Gauthier.FANMUY@3ds.com

A. Durantin
e-mail: Arnaud.DURANTIN@3ds.com

S. Miet · V. Pegon
Dassault Systèmes – Design Studio, Vélizy-Villacoublay, France
e-mail: Segolene.MIET@3ds.com

V. Pegon
e-mail: Valerie.PEGON@3ds.com

G. Fanmuy et al. (eds.), *Complex Systems Design & Management*,
DOI 10.1007/978-3-319-49103-5_4

1 Introduction and Scope

1.1 Why This Paper

The goal of this paper is to report on a dialogue and experimentations between design thinkers and systems engineers, to improve both practices through each other's approaches. This journey started almost three years ago through personal contact and curiosity towards each other. We soon came to the conclusion that we had similar perspectives on some aspects. The key revelation about the importance of this topic happened during an industrial project. The Design Studio works closely with companies from different industries, helping them to transform their innovation process through methodologies borrowed from design thinking. By doing so, we explore and test approaches that include methods, tools, and demonstrators. During this project, we understood the current situation and the needs we will explain in this article.

1.2 Focus

Our focus will be on the early stages of innovation, also called "operational phase", when companies identify the opportunities they may want to pursue and define their product and/or service propositions. We will not develop the following phases of design development. We will also address complex systems containing products and services. This heterogeneity is more and more common and very difficult to manage—hence the need for new approaches.

1.3 Design Thinking

What is Design Thinking

Design thinking is an approach that encompasses project organisation, posture, methodologies and tools, used by companies to create new products and services. The growing complexity of businesses and their systems requires a flexible way to explore desirability (what people want and value), feasibility (what is technically possible) and viability (how it could be a successful business) [1, 2]. To do so, design thinking stands out through user empathy, collaboration and iteration (using prototypes).

The design thinking methodology

While the ambitions of design thinking are often overemphasised, its results are as good as the people practicing it and the context it lives in [3]. It is by no means a magic recipe, where a process can be applied to reach amazing results.

Understanding

One of its key strengths is to enable space and time to question the brief. This is the first phase, Understanding, where digging deep behind the challenge is essential to success. To do so, design thinkers need to explore the relevant topics with a wide lens, covering social, technology, business and identity. Several methods are combined: desk research, trend research (critical for longer term innovation), qualitative user research (necessary for short term innovation) and expert interviews. They are completed by a creative, collaborative and diverging phase to imagine opportunities. The result is a clear brief, objectives and success criteria that will drive the next phases.

Definition

It is then time to define a strategy; i.e. how to tackle a specific opportunity. In this phase, Definition, collaboration is key to lead to propositions desirable, viable and feasible. At our Design Studio, we use creative one to two-day sessions mixing a diversity of profiles, from within the company and outside. The challenge is to move from knowledge to new ideas. This requires a creative process that cannot be controlled but can be facilitated with inspiration, rhythm, state of mind and a methodology to move on step by step towards concrete scenarios of projected futures.

Conception

The next phase, Conception, is getting closer to the industrial design process, where teams work on more focused user research, creative sessions with many design proposals (a second phase of divergence), early mock-ups, tests and yet more iteration. For a robust design thinking approach, this phase must also be done in a collaborative way, ensuring coherence between the different elements of the puzzle.

Development, production & promotion

The last phases, Development, Production and Promotion, are continuing on the same approach: collaboration and iteration. Because they are not the focus of this paper, we will not develop them.

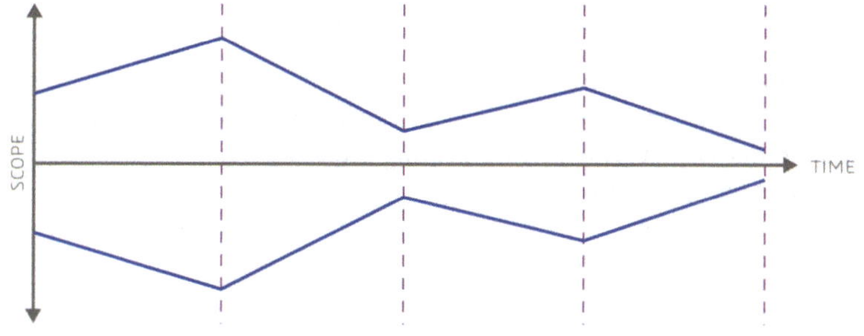

The design thinking alternates divergence and convergence.

Design thinking representations

The *raison d'être* of design thinking is innovation, to propose new products and services that can bring value to the market. Design thinking can serve both disruptive innovation—by focusing on foresight and longer-term vision—and incremental innovation—by focusing on current usage and context. To communicate ideas for future user experiences, the Design Studio uses storytelling tools such as user journeys, scenarios, storyboards, movies and demonstrators. These tools describe elements of a system from users perspectives.

1.4 Systems Engineering

What is systems engineering

Systems engineering is an interdisciplinary collaborative approach meant to enable the realisation of successful systems by considering its complete lifecycle. It is based on the concept of "a system": "an interacting combination of elements to accomplish a defined objective" [4]. In order to manage complexity, it organises project datasets into several levels of abstraction from a general overview of the problem to solve, to the most concrete and detailed description of the system.

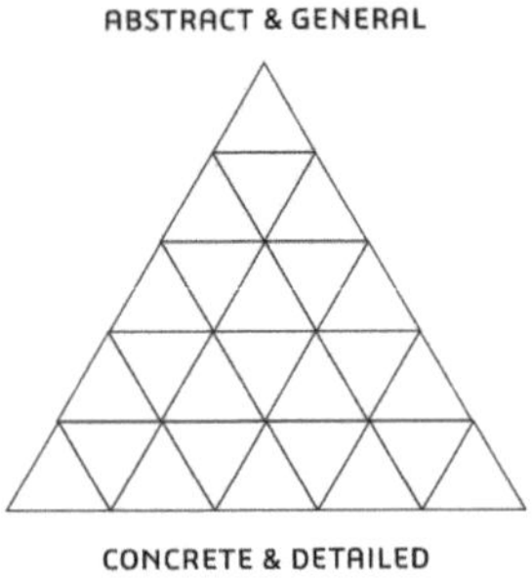

Systems engineering methodology

There are various methodologies used in industry, such as [5]:

- IBM Harmony for systems engineering
- INCOSE Object-Oriented systems engineering Method (OOSEM)
- Vitech Model-Based System Engineering Methodology
- JPL State Analysis (SA)
- Cofluent methodology
- CESAMES Matrix methodology

Dassault Systèmes has developed its own methodology—Modelling Methodology for Systems—based on state-of-the-art systems engineering practices, to address all disciplines involved in engineering a system. The table below summarises its structure, and is followed by its principles.

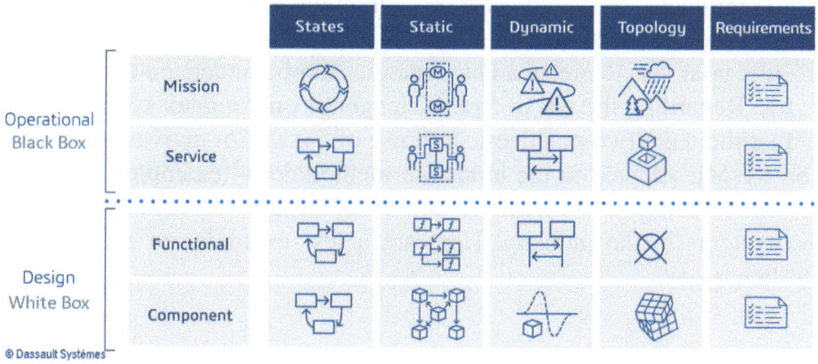

- Separation of the outside (black box) and the inside (white box) of a system.
- Management of layers of abstraction, integrating life cycles and business objects.
- At each layer, a set of views for different perspectives, ensuring a consistent and complete definition with states and modes, architectures and contexts (static), scenarios (dynamic), physical environments (topology) and requirements.

A language to support the methodology

When modelling systems, it is necessary to use a language that enables all the people involved to understand each other. Several languages exist, depending on what is expected at any given step and for certain areas of the system.

- **Semantic languages** (e.g. English): the main type of language used to describe a system in documents such as specifications or design descriptions.
- **Modelica**: used to create and simulate multi-physics models
- **UML (Unified Modelling Language)**: describes a software behaviour with a set of diagrams, and **SysML (SYStem Modelling Language)**: derived from UML and adapted to systems
- **FFBD (Enhanced Function Flow Block Diagram)**: a systems engineering model representing the system's behaviour.
- **IDEF (Integration DEFinition):** family of modeling languages in systems and software engineering, that covers a wide range of uses, including functional modeling, simulation, object-oriented design and knowledge acquisition.

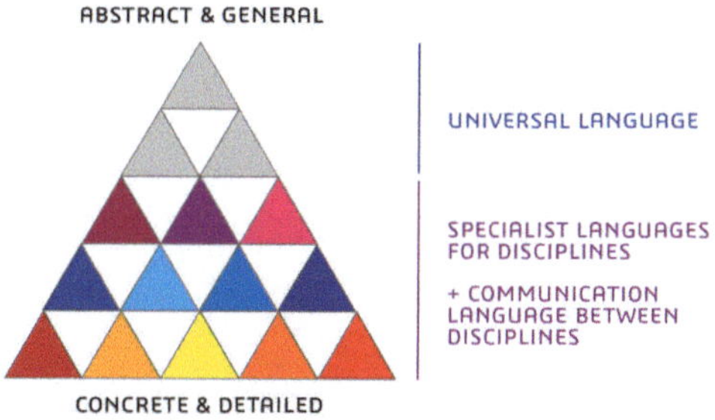

Language needs depending on level of concretisation

Natural languages are widely used for communication but, because of its variety, they often lead to misunderstandings. In Model Based Systems Engineering (MBSE), the goal was to create a language that can be understood in only one way: SysML [6]. Regarding the use of SysML language on complex system projects, we can say that there are two practices: strict use of SysML or derivation of a language based on SysML adapted to the industrial methodology (example: Thales with the Arcadia methodology). Nowadays, these representations start being used to describe systems in the industry. Here are a few examples of representations in different languages:

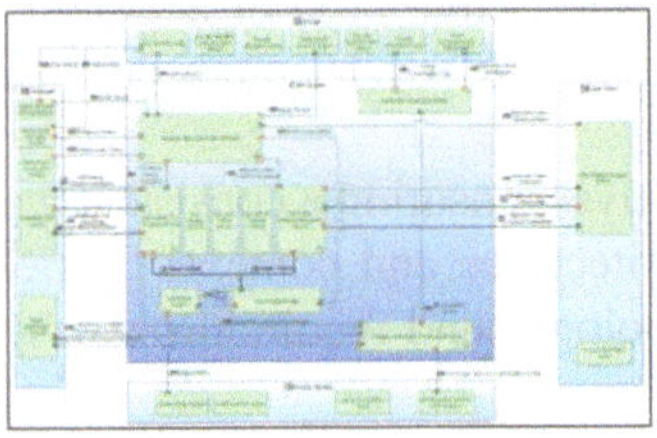

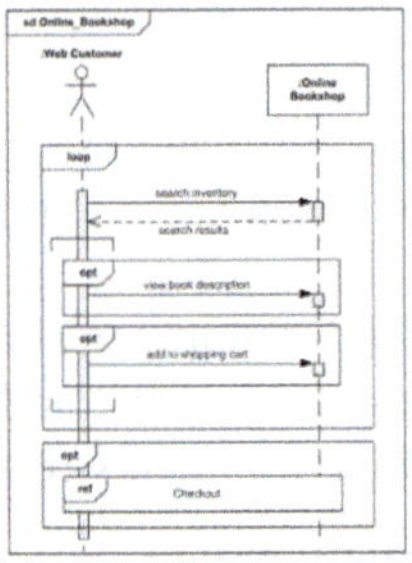

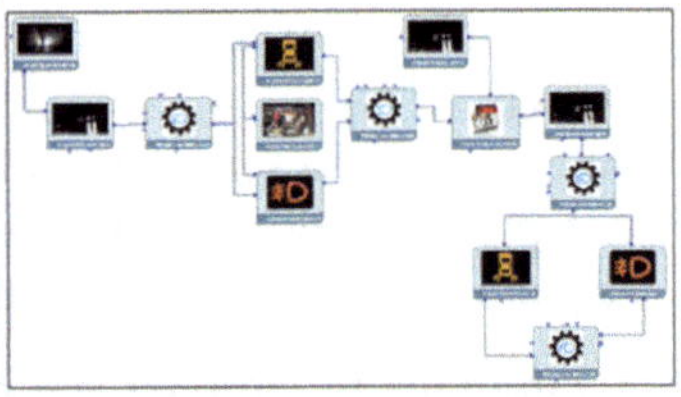

Thales Melody *IBM Harmony* *Dassault Systèmes MMS©*

- *System overview: models the system's context and perimeter, with relevant stakeholders, effective interactions and interfaces with the system.*
- *Sequence diagrams: a set of models showing the sequence of actions between stakeholders and system.*
- *Functional chain: for a service, the functional chain describes the interactions between internal functions*

These diagrams provide a complete description of the system. However, many project stakeholders see them as a language specific to systems and electronic architects, difficult to read and understand. We believe this point is nowadays a limit to the development of MBSE in the industry. On the chart below, you can see that these kinds of representations cover less than 25 % of project's stakeholders in industrial engineering projects. It is clear that these representations are far from covering the real needs. They enable experts to design the system but the result of the design is not understandable to the majority.

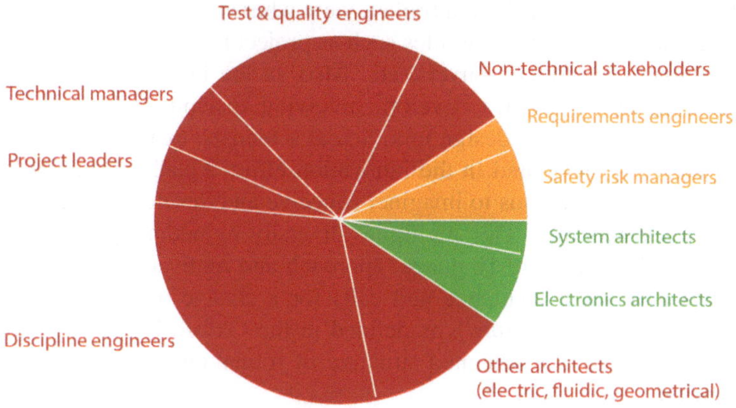

Level of SysML legibility in a project

2 Insights from our comparison

2.1 Insights

After explaining each other our approach and testing with a few sample topics, we started to identify common drivers that are leading us closer together.

Diversity of people

As mentioned in the introduction, complex systems made of products and services are getting more common. Our industrial experience has shown us that currently, around 40 disciplines are involved in complex systems designs, such as a car, a rocket or an airplane. Also, many industrial companies offer services rather than just selling products. For example, the business of Rolls Royce is to sell flight hours, not jet engines. This means new disciplines must be brought to the innovation process, to conceive products and services together, as a coherent whole, as a system of systems. However, these disciplines are not all from the engineering family. In addition, the increased complexity and heterogeneity in companies' offers also make it harder to make decisions. For example, board members may have a robust experience in a few areas, but they can't be experts in all disciplines involved in complex projects. They need a clear overview and means to explore the systems their teams are working on, without the burden of deciphering an unfamiliar language and without a broken view where the whole is not legible. The same applies to other wide scope roles such as project managers, product managers, service designers, marketers, buyers, etc. Also, in the last two decades, we have seen user centred approaches develop, involving disciplines beyond traditional engineering, marketing and design mix. Social scientists, experience designers and even end users are integrated in the innovation effort. Finally, companies need to work together, in ecosystems to imagine, conceive and develop future systems that go beyond their expertise. This multiplies diversity by adding a corporate culture layer; hence a divergence of language, approach and motivation. Linking all these perspectives is a major challenge and calls for a change of paradigm in design, opening an era of "architecture", as defined in the Oxford Dictionary of English: "the complex or carefully designed structure of something."

Disruptions

Most industries are seeing deep disruptions as the digitalisation of business is going forward. Many old truths are replaced by new ones. The taxis' stronghold is threatened by Uber whose model will probably be disrupted again by new offers made possible by technologies such as autonomous cars and block chain. These radical and deep changes make it very hard for established companies to adapt. They have invested years and millions in specific sets of competences, technologies and other assets. Yet, these can become irrelevant. One option is to wait and see, the other is to rebuild. Walmart has chosen this second route, as it plans to close 269 stores in 2016 and open 405 new ones to adapt their offer to the new context. In this kind of dynamic context, questioning a company's offer has become a priority. Many companies are turning to design thinking, among other approaches, to define a vision on their potential future and imagine new systems.

Interaction with computers

As described above, the current modelling languages for engineering systems are not universal and are understood by a small minority only. In addition, they haven't been optimised for human legibility, but for computer readability. For example, texts are often small, there is no visual hierarchy and content is in black on white. In the last ten years, trends in user interface and data visualisation have given us innovative alternatives with visual representations and interactivity. Our personal interfaces have become visually easier to read and have influenced most professionals who are starting to expect this level at work too. Designers and platforms such as Gephi have explored visually clear and appealing ways to represent complex data.

Co-authorship network of 8,500 doctors and scientists publishing on hepatitis C virus. Data from Medline, processed using Python, and visualised with Gephi. Creative Commons licence.

However, many professional tools haven't harnessed these new possibilities. Users still need to adapt to them. For example, sequence diagrams (SysML) are describing events in a way that is easy to read by computers, but totally unappealing and difficult to read, at a cognitive level, for fellow humans. It is dangerous to think this is superficial matter. We conducted interviews with a range of professionals in different technical disciplines who identified visualisation as a problem. If something is not appealing, nobody will make the effort to read it. Visual models need to be at the same time appealing and easy to read. The improved ability of computers to decipher our content (image recognition, semantic analysis, etc.) and to interact with humans (through voice, movement, touch, etc.), make today the perfect time to revisit these representations.

Human complexity

There is something else at stake—a need for motivation to work together in collaborative ways. People from different disciplines and cultures value different things and are motivated differently. In particular, emotions can get in the way or tremendously boost imagination, team productivity and happiness. Emotions cannot be controlled, but marketing and science have explored this area well enough to teach us that perception through all our senses are critical to nudge us towards a "desired" mood. Providing a refreshing pause in day-to-day operations to think about the future, is often cited as the best part of the design thinking workshops we organise at the Design Studio. These moments enable people to build a team, a community that can last beyond these sessions and that has learned to work together. Besides, humans with all their senses and behaviours need to be considered as part of the system, not only as elements interacting with the system.

Shared ambitions

Systems engineering and design thinking share a few intentions; mainly a collaborative and holistic approach. It seems that systems engineering has reached a plateau in regards to teamwork, and needs to reassess its approach and tools. This is one of the key domains where design thinking can bring value, as we will see in the following chapter. Equally, design thinking has limits when it comes to being holistic. While it performs well at balancing desirability, feasibility, viability and identity, it doesn't achieve full exhaustiveness. For example, some life cycles, contexts or stakeholders won't be considered, while a strong focus will be made on others. Design thinking is rarely neutral, but adopts a posture that introduces a bias. Without losing this unique perspective, systems engineering can bring great value by organising information and helping reach more exhaustiveness. It will also enable to make a bridge for a continuous engineering between innovation and development phases. The visual below shows how the definition of a system moves forward during the innovation process.

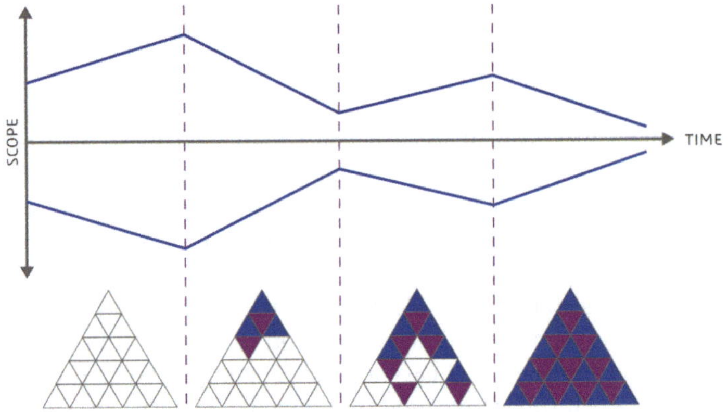

2.2 *Strengths and Weaknesses*

With all these insights in mind, we noticed that, often, strength in one discipline could be balanced by a weakness in the other, and vice versa. The table below summarises this comparison.

	Strengths	Weaknesses
Design thinking	Universal language and visual representations (describing human experiences) Collaborative methodology made for accessibility Exploration methodology enabling divergence Methodology to integrate human complexity Capacity to fail fast and iterate Holistic vision (e)	Not very structured, lack of standards Not studied in robust academic context (a) Blind spots outside end user experience, "in use" life cycle and limited contexts of use (b) Lack of visual representation standards

(continued)

(continued)

	Strengths	Weaknesses
System engineering	Highly structured approach with a precise vocabulary (c) Science practiced academically Systemic and exhaustive (stakeholders, contexts, life cycles…) Standard visual representations Exhaustivity with a framework (e)	Lack of accessibility for all the disciplines (c) Lack of team working methodologies (d) Lack of exploration methodology Lack of consideration for humans beyond actions and basic ergonomics Time consuming (modelling)

Notes

(a): Design thinking emerged recently (in the 1980s) and is not widely studied in an academic context. While design is much older, it has been mostly considered a practice not suitable for research. However, design research has started to take off in the last ten years.

(b): Design thinkers focus largely on the end users and often forget about the other stakeholders who also need a positive experience to satisfy the end users. The service design discipline, close to design thinking, has focused efforts on solving this issue in the last ten years, adopting a systemic approach.

(c): The vocabulary used in systems engineering is standardised, very precise and specific. This is a double-edged sword: people in the know—fellow systems and software engineers—understand it around the globe; while others see it as a foreign language, as is the case SysML (understood and accepted by a small portion of stakeholders, as described above). Learning a language without daily practice is extremely difficult, so forcing a minority's language onto the majority is probably not the most realistic approach.

(d): We consider systems engineering lacks methodology to work in teams, as a patchwork of humans. Its only mean is sharing structured information. For example, there is no methodology to generate information such as stakeholders, contexts, use cases, etc.

(e) It is very interesting to note the difference in achieving a holistic approach. Systems engineering provides a structured framework for many people to structure the information within, looking for exhaustiveness. On the other side, design thinkers imagine (with others) a coherent whole, in one picture. However incomplete and unstructured as it may be, this picture of the whole communicates a vision to reach - as opposed to boxes to fill.

3 Experiments

After comparing each other's practice at a theoretical level, we did a few internal experiments along with customer workshops and deliverables to test and practice how the other team would work at specific stages of projects. In doing so, we ran into difficulties that helped us build the following hypotheses.

3.1 Timing

If there is one insight to remember when trying to combine our approaches, it is the difference in time. Design thinking is most useful when approaching "wicked problems" [7]. This means, there is a set of challenges, often intertwined and ill defined, that need a fresh look at. This is the core of our design thinking work at the Design Studio, where we answer open questions with no focus on products (physical of digital). It can often be summarised as "how can we reinvent our industry sector?" The result of our approach is an incomplete but coherent picture of a desired future. In the industrial context, systems engineering very rarely starts from a blank sheet of paper and opening up to challenges outside the system. More importantly, its capacity to model and simulate a system requires a certain level of stability and top-level definition. It is most powerful at supporting convergence, but doesn't have means to deal with divergence phases where information is generated. For these reasons, systems engineering naturally comes after the first phase (divergence) of design thinking, to structure the system that has started to emerge. Later in the process, during the second divergence phase, systems engineering can go on near stand-by, waiting for elements to be defined. This is exactly how it happened on one of our recent design innovation project. The Design Studio helped the client create scenarios for a future product and service system. After this creative and storytelling phase, the design thinkers worked with systems engineers to structure missions, stakeholders and services, giving a first structured view to the client and supporting their decision making process. The illustration below shows the involvement of each discipline along the innovation process.

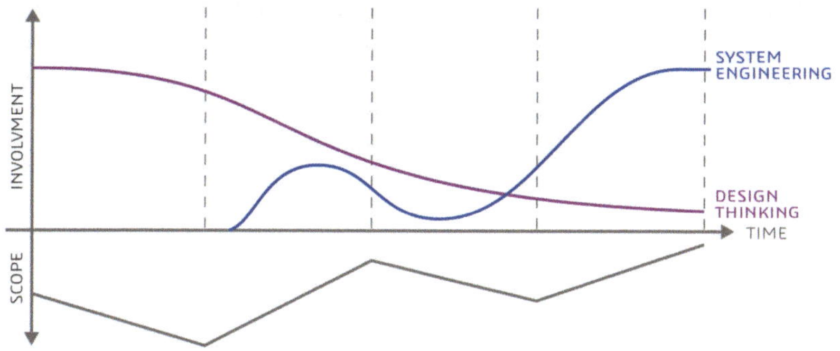

3.2 *Caution*

Culture

Each discipline has its own culture and ways of working. Designers and design thinkers suggest a lot of ideas, iterate, throw them away, turn them upside down, give them to someone else, etc. At some point, they need to select one to develop, but they will continue to iterate significantly for quite some time. They even go through a second divergence phase in the development phase. Engineering is focused on finding the right answer to study closely and prove right. However, systems engineers try to keep the possibilities open as late as possible to avoid limiting possibilities. For example, rather than defining the system as a specific robot (solution), they abstract a "System Under Development"—SUD. This leaves the possibility for expert disciplines to fill in the blanks. On the other side, design thinkers leverage the evocative power of imaginaries (evolving sets of narratives and forms [8]) to describe early ideas. While engineers speak about the SUD—an immensely neutral expression—design thinkers are more expressive by speaking of a compass, for example. This value-oriented approach helps kicking off creativity and alternative solutions. In the client project mentioned in 3.1, early component concepts helped the client identify a wider set of services and business opportunities. This difference is easily understood when we consider the ultimate aim of each discipline. Design thinking provides and describes a vision to reach, while systems engineering provides a structure to define and link many elements. These are two different ways to work we don't believe should be muddled together. Their differences bring richness in a combined approach.

Vocabulary and approaches

Another point of difference related to culture, is the differing ways of "breaking down" information into categories, exact terms and points of view. This is best explained through an example. We imagined the following scenario: the early design phase of a system to remove mines in busy touristic areas in Cambodia. Both teams, design thinkers and systems engineers, agreed in the value of identifying the stakeholders, but didn't agree on who they should be. Design thinkers added dogs and dog masters to the list of stakeholders, while systems engineers had listed dogs in the components because they are part of the system to develop. For design thinkers, the stakeholders can be equally inside or around the system.

3.3 Combination

Our design thinking process primarily takes place before the systems engineering method. The design thinking methodology helps imagine the system, whereas systems engineering grounds it. Both methodologies can act side by side and turn out to be complementary. The design thinking approach performs well in leading and expanding divergent thinking, but it lacks tools to collect and organise the results in a real content structure. In contrast, systems engineering benefits from a structuration of content but doesn't have the means to diverge - a barrier to include more disparate and alternative content. However, different steps in these methodologies match and present the same types of content, sometimes at different scales. For example:

- In systems engineering, context diagrams presenting the stakeholders and their interface with the system is close to the design thinking system overview showing the system and the ecosystem around it.
- In systems engineering, missions and service scenarios are close to the synopsis and scenarios in design thinking. All describe the actions between the stakeholders and the system.

Below is a summary of how our processes run in parallel, starting in the first phase of convergence. It is important to note that both processes are iterative within each phase. The scenarios and the first elements of the design thinking phase 2 correspond to a pivot moment as they communicate a non-exhaustive set of system elements. From this, systems engineers can start structuring information into a robust model. In the client project mentioned earlier, we represented, in the same model, the "traditional" elements listed above in 3.1 (missions, stakeholders and services) and two types of information from the design discipline: experiences lived by specific stakeholders in the scenario and component opportunities. These bring a point of view on how to design the system and start to form briefs and specifications for the project.

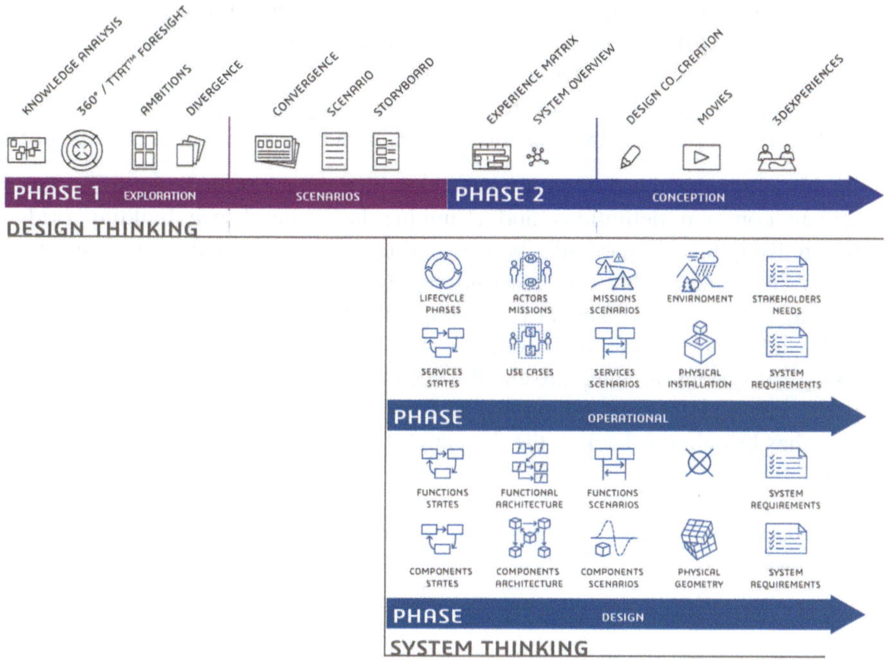

Overview of the design thinking activities and systems engineering views

4 Conclusion

Edgar Morin said complex thinking is aimed at "encompassing rather than separating, linking rather than segmenting" [9] (our translation). Design thinking imagines the whole in a global coherence, accelerating innovation with illustrations of a few potential bits of what we could expect to become. Systems engineering models and simulates products (all the bits and links) to make it happen. To make Edgar Morin's vision a reality in innovation, both design thinkers and systems engineers have to work together, with combined methodologies and shared elements of language. Following our experimentation, we believe it is also important to develop tools that link both approaches. Systems engineering has two major needs: modelling systems architectures in an accessible and ergonomic way, and helping convergence. Equally, design thinking also has two critical needs: tools to accelerate the creation of deliverables and structuring content to facilitate hand-over. What if there was a "repository" that provided continuity from design innovation to engineering? We believe it could serve different kinds of profiles, which we organise in four categories:

- Integrators: people who link disciplines and elements of systems (e.g. project managers, experience designers, product managers)
- Approvers: people who decide (e.g. CEOs)
- Contributors: people who create (e.g. systems engineers, product designers)
- Advisers: people who are consulted (e.g. legal advisers, technical experts, users)

In this adventure ahead of us, we have identified three key challenges. The first is to create common definitions and a unique language; hence building bridges between disciplines, in particular between systems engineers and novices. Next, we need to create accessible and relevant representations for different levels, from macro to expert views. The last challenge is to define representations that are actually adequate in the real context of use; hence understanding the actual workings of systems architecture and the needs of different users, not compromising experts' views. However, ultimately, the main stretch is to combine contrasting approaches bringing us back to the meaning of "technology": the alliance of art and science.

Aknowledgements Anne Asensio, Etienne Boulanger, Joseph Cavé, Stan Edde, Morgan Lartigue and our clients

References

1. Brown, T.: Change by Design (2009)
2. Kolko, J.: Design Thinking Comes of Age, HBR September 2015
3. McCullagh, K.: Beyond Design Thinking, Plan
4. ISO/IEC 24765 Systems and software vocabulary
5. INCOSE-TD-2007–003-02—Survey of Model-Based Systems Engineering (MBSE) Methodologies
6. OMG Systems Modeling language - http://www.omgsysml.org
7. Rittel, H., Webber, M.M.: 1973 characteristics of wicked problems
8. Chaire Des Imaginaires, TelecomParisTech & Rennes 2 University, manifesto
9. Morin, E., L'An I de l''ère écologique et dialogue avec Nicolas Hulot, p. 107, Tallandier (2007)

Validation of Industrial Cyber-Physical Systems: An Application to HVAC Systems

Thao Dang, Alie El-Din Mady, Menouer Boubekeur, Rajesh Kumar and Mark Moulin

Abstract We describe a validation approach for Simulink models of industrial cyber-physical systems (CPS), based on an adaptation of a coverage-guided test generation method for hybrid systems. Modelling an industrial CPS requires integrating heterogeneous components, which introduces high complexity in model verification. Using Simulink, which has become a de-facto industrial tool, heterogeneity comes from combining different formalisms (Simulink blocks, Stateflow diagrams, Matlab and C functions, etc.) and mixing different types of dynamics (discrete, continuous). Since the interactions between such components are often too complex to be faithfully captured in an existing mathematical modelling paradigm, we resort to treating them as black box systems while trying to exploit as much as possible a-priori knowledge about them. We first describe our approach: extracting from a Simulink model the information to define the main ingredients of the test generation framework, in particular environment inputs in which faults could be injected and critical states that require good coverage. We then illustrate the approach with an industrial model of an HVAC (Heating, Ventilation and Air Conditioning) system.

This work was supported by United Technologies Research Center, Ireland.

T. Dang (✉)
VERIMAG/CNRS, Grenoble, France
e-mail: thao.dang@imag.fr

A.E.-D. Mady (✉) · M. Boubekeur (✉) · R. Kumar (✉) · M. Moulin (✉)
United Technologies Research Center (UTRC), East Hartford, USA
e-mail: MadyAA@utrc.utc.com

M. Boubekeur
e-mail: boubeKM@utrc.utc.com

R. Kumar
e-mail: KumarR@utrc.utc.com

M. Moulin
e-mail: moulinm@utrc.utc.com

G. Fanmuy et al. (eds.), *Complex Systems Design & Management*,
DOI 10.1007/978-3-319-49103-5_5

1 Introduction

Cyber-physical systems (CPS) design is an emerging domain which has rapidly grown in terms of methodologies and applications. Roughly speaking, cyber-physical systems are integrations of computation with physical processes. They are often heterogeneous systems admitting components of different types of dynamics (continuous and discrete), specified using different mathematical models. Assuring correct behaviours of CPS is crucial for safety-critical applications. Due to the complexity of formal verification (which is based on exhaustive analysis and thus is limited to applications for low-dimensional systems), alternative approaches, which can be applied to real-life high dimensional systems, are very desirable. One of these is testing which is the validation technique par excellence in industrial practice. Our goal is thus to adapt a hybrid systems testing technology [5], which uses a coverage measure and can generate test stimuli that allow a good coverage of the behaviours of interest. Since this technology was developed for hybrid automata (a mathematical model for describing systems with mixed continuous and discrete dynamics), it must first be adapted to the specific features of industrial models. We focus on Matlab/Simulink, which is a standard tool for modelling, developing and testing industrial CPS. In particular, we demonstrate this approach with an HVAC (Heating, Ventilation and Air Conditioning) model constructed using Simulink. HVAC systems are used to supply a thermal power to a thermal demand in a building in order to respect the user comfort requirements, e.g. CO_2 level, temperature, etc. There are two main types of HVAC, air-based and hydronic, where air-based HVAC uses the air flow as the main medium to transfer the thermal load from the source to the demand. Air-based HVAC is the most known industrial HVAC system; therefore we consider a case-study of air-based HVAC for a real industrial demo-site. The Simulink model of this HVAC system exhibits many features of CPS, in particular a combination of physical processes and computation of control laws, described using a mixture of hierarchical modelling formalisms (Simulink discrete and continuous blocks, lookup tables, embedded Matlab code). The complexity of this model allows us to investigate the difficulties in exporting model-based testing techniques, often developed for some specific classes of mathematical models, to an industrial design environment such as Simulink. Indeed, the lack of formal semantics for Simulink models constitutes a major obstacle in applying formal analysis techniques to Simulink industrial models, since these techniques require an effective mathematical description (e.g., some closed-form analytic representation of dynamics) to be available. A number of approaches have been proposed to generate automatically a mathematical description from a subset of Simulink blocks (the semantics of which can be formally determined) with restrictions in their connection (for example in [17] this subset does not include continuous-time dynamical blocks and zero delay feedback is not allowed). To resolve this semantics issue, we resort to considering the behaviours of the Simulink simulator when it compiles and 'interprets' Simulink models and we do not seek a formal definition of semantics (as in [1, 3, 17]). Note that such formal semantics (which could be defined for a subset of Simulink blocks) allow comparing numerical simulation traces and mathematical behaviours, which is important for the approaches such as model checking, abstract interpretation. Our goal is rather testing, for which we could use simulated behaviours as the "semantics" intended by designers when they construct their models.

The paper is organized as follows. We first show how to fit Simulink models in a formal testing framework. We then show how to adapt the coverage-guided test generation techniques hybrid systems [5] for Simulink models. We illustrate this result with the HVAC model treated using an implementation, which integrates these adaptations in the test generation tool HTG [5]. Before continuing, we discuss related work on coverage measure and test generation for Simulink models.

Related Work Test coverage metrics have been used by a number of testing tools for Simulink models, among which we can mention Reactis Tester,[1] T-VEC Tester,[2] REDIRECT [15]. These coverages are nevertheless mainly structural coverages, such as Statement Coverage, Decision Coverage, Modified Condition Decision Coverage (MC/DC), for which data flow and control flow are two main criteria; essentially the former measure the flow of data between variable updates and references to the variables, while the latter measure the flow of control between statements. Test coverage in terms of the number of tested configurations was proposed in [2] where the configurability or variability of both the system and the test architecture is described by feature models, from which Simulink models are generated for test purposes. The major difference between these coverage measures and the measure we use in this work is that these coverages are defined on syntactical descriptions and measured in terms of statement executed, Boolean expressions evaluated, configurations tested, *etc.*, while our measure, defined on the set of temporal evolutions of the system under test, is thus more appropriate to handle dense-time temporal properties.

While test coverages provide assurance that important or representative behaviours are tested, another approach, implemented in the tool S-Taliro [10] and Breach [7], seek the worst case behaviours using a notion of robustness which describes how close the system is to the satisfaction or violation of a property. This problem is then formulated as minimizing the robustness over the input space and all the possible initial conditions. This approach however either requires knowing the simulation function which generates the traces of the system, or uses global optimization based on local search methods. The robustness-based approaches can be seen as complementary to our coverage-based approach, since the former try to find a worst case behaviour while the latter tries to cover well all possible behaviours. When the former cannot find an erroneous behaviour due to the limitation of global optimization algorithms and this observed error absence cannot be used as a formal correctness proof; in this case a good coverage would be desirable to enhance the confidence in the result. In [9], the algorithms of the HTG tool have recently extended to include robustness metrics with respect to properties specified using STL (Signal Temporal Logic) [8].

[1] http://www.reactive-systems.com/simulink-testing-validation.html.

[2] https://www.t-vec.com/solutions/simulink.php.

2 Framework for Testing Simulink Models

Testing is mainly concerned with finding an observable behaviour that is different from what is expected. Considering a Simulink model as a system under test, the behaviour in question is a simulation trace produced by the Simulink simulator. However, an essential difference between our testing approach and the usual simulation approach is that our approach produces a tree of simulation traces rather than a set of single simulation traces. Indeed, in each simulation step, the exploration can be continued from a previously visited state, and not necessarily from the current state. As we shall see, this ability is important for achieving a good test coverage. Our concrete goal is thus to develop a program, called "tester", that interacts with the Simulink simulator to guide the simulation process. To this end, we need a formal framework to describe the mapping from the input signals to the output signals produced by the simulator.

The Simulink simulator can, for a given input signal, returns the corresponding output signal. It is however important to have access to the evolution of the internal variables of the model and the simulator in order to guide the simulation towards an erroneous behaviour and to assure a degree of test coverage. Using the behaviour of the simulator to define the semantics of Simulink models is a commonly accepted approach [3, 12]. It is important to note that the behaviour of the Simulink simulator is deterministic in the sense that to run a simulation, the initial conditions of all the Simulink blocks must be defined. If the model has inputs (coming from the external environment), a particular function over time must be defined for each input. If a model parameter is not specified, it takes its default value. Similarly, a solver can be specified, together with all the simulation parameters. Simulink models could admit random noises, and the associated noise generators should also be defined. The simulator then simulates the model under this specific configuration, and each simulation run produces a single simulation trace. Since the simulator uses numerical integration that can determine the values of the variables only at discrete time points, we represent the signals involved in the resolution process by sequences of time points coupled with multi-domain signal values (describing physical quantities), for example $\mathbf{x} = (t_0, \mathbf{x}(t_0)), (t_1, \mathbf{x}(t_1)), \ldots$ where the time points $t_k \in \mathbb{R}_+$ for all k (where $\mathbb{R}_+$ denotes the set of positive real numbers). We remark that to approximate the solution at discrete time points, the Simulink simulator uses advanced continuous-time resolution algorithms to handle interaction between continuous and discrete dynamics (specified by differential algebraic and logical constraints). This signal representation however cannot express explicitly discontinuities in the solution. More complex signal representations that unambiguously capture discontinuities are also used, such as in [12].

We use the following non-autonomous discrete-time dynamical system model to describe the behaviour of the Simulink simulator. A dynamical system $\mathcal{M}$ is a tuple $(\mathcal{X}, \mathcal{U}, \mathcal{Y}, X_0, \mathcal{F}, \mathcal{H})$, where $\mathcal{X}$ is a set of state values and is called the state space, $\mathcal{U}$ is a set of input values, $\mathcal{Y}$ is a set of output values; $\mathbf{x}$ (*state variables*), $\mathbf{u}$ (*input variables*) and $\mathbf{y}$ (*output variables*) denote functions mapping a time point t to a state value $\mathbf{x}(t)$ in $\mathcal{X}$, to an input value $\mathbf{u}(t) \in \mathcal{U}$, and to an output value $\mathbf{y}(t) \in \mathcal{Y}$ respectively, X_0 is the set of initial states $\mathbf{x}(t_0)$.

$$\mathbf{x}(t_{k+1}) = \mathcal{F}(\mathbf{x}(t_k), \mathbf{u}(t_k), h_k), \mathbf{x}(t_0) \in X_0 \quad (1)$$

$$\mathbf{y}(t_k) \quad = \mathcal{H}(\mathbf{x}(t_k), \mathbf{u}(t_k), h_k). \quad (2)$$

A *trace* of $\mathcal{M}$ is a sequence of $\left(t_0, \mathbf{x}(t_0), \mathbf{u}(t_0), \mathbf{y}(t_0)\right), \left(t_1, \mathbf{x}(t_1), \mathbf{u}(t_1), \mathbf{y}(t_1)\right), \ldots$. In general, $\mathcal{X}$, $\mathcal{U}$ and $\mathcal{Y}$ can be a product of different domains (such as the Boolean, integer, real domains) which are admissible by Simulink. Also, the durations h_k should not be confused with the internal time steps of the simulation algorithms. Note that the above discrete-time model can only be seen as an abstraction of the Simulink simulator. As mentioned earlier, Simulink models of CPS operate in continuous time, thus the input signals must be defined in continuous time. In this work, we associate with a sequence of input values a piecewise continuous signal which is to be fed to the simulator, for example the signal corresponding to a sequence $\mathbf{u} = (t_0, \mathbf{u}(t_0)), (t_1, \mathbf{u}(t_1)), \ldots$ is $\tilde{u}$ such that $\tilde{u}(t) = \mathbf{u}(t_k)$ for all $t \in [t_k, t_{k+1})$. It is possible to consider other classes of input signals, for example piecewise linear signals, by linearly interpolating two consecutive input values. In the remainder of the paper, for simplicity of notation, we sometimes write $\mathbf{x}$, $\mathbf{u}$, $\mathbf{y}$ to denote the values of these functions, omitting the associated time points.

We now discuss how to fit the syntactic description of a Simulink model in the above high-level description formalism and the testing framework. Given a Simulink model S, let $\mathcal{M}_S = (\mathcal{X}, \mathcal{U}, \mathcal{Y}, X_0, \mathcal{F}, \mathcal{H})$ be the dynamical system modelling the behaviour of the Simulink simulator on S. The components of $\mathcal{M}_S$ are defined as follows.

Input signals. In principle, before each simulation run, all the inputs of a Simulink model must be fully defined for the whole time horizon of interest. In our testing approach, the input signals, in contrast, can be dynamically defined during each simulation run guided by the tester. We call the input variables that the tester manipulates *control inputs*.

Output signals. The outputs $\mathbf{y}$ correspond to the signals in the model we want to observe. This signals often involve the property to test.

Dynamics. The functions $\mathcal{F}$ and $\mathcal{H}$ model the behaviour of the simulator executed on a Simulink model when all the simulation options are fixed. We do not require the functions $\mathcal{F}$ and $\mathcal{H}$ to be known in an analytic or symbolic form. They could be complex, combining discrete and continuous dynamics with algebraic constraints via feedback loops without delays.

State variables. Our approach allows a dynamical exploration, that is the tester can decide to continue the simulation from a previously visited state (and not necessarily from the current state), it is thus necessary to reinitialize the internal variables of the simulator. It is thus important to have the ability to identify and manipulate the state variables of the dynamical system modelling the behaviour the simulator. The question of terminology is important here. The notion of "state" in Simulink is not formally defined. Indeed, when the information about a Simulink model is printed out, only its explicit state variables are reported in some syntactical description. We call such state variables *controllable state variables*, because they can be accessed and controlled by the tester. For example, the Simulink simulator treats the Integrator block as a continuous-time dynamical system with one state variable $x \in \mathbb{R}$, one input variable $u \in \mathbb{R}$, one output variable $y \in \mathbb{R}$: $\dot{x}(t) = u(t), x(0) = x_0$ and $y(t) = x(t)$, where the x_0 is the initial condition of the block. Using our discrete-time dynamical system model to describe the behaviour of the Simulink simulator, given a time point t_k and a time step h_k, the function $\mathcal{F}$, computed by the simulator, gives the value of $x(t_{k+1})$, with $t_{k+1} = t_k + h_k$, by integrating the input signal. However, there are internal (hidden) states that are not reported, and the user cannot have access to such internal information. It is also possible that industrial Simulink models contain blocks (such as continuous-time delays) and Matlab code which essentially represent systems with an infinite number of state variables. As an example, the HVAC model, contains many blocks of heterogeneous nature that also exhibit this feature. Such Simulink components constitute a challenge for model-based analysis techniques that perform state space exploration. While we can assume that Simulink is capable of simulating a complex system to a sufficient level of accuracy, we cannot assume that from the Simulink simulation traces it is possible to reconstruct all the hidden states. A specific solution to address the difficulty in identifying state variables will be described in Sect. 3.

Coverage Guided Test Generation In the remainder of this section, we assume that all the state variables of the simulator are controllable by the tester. The simulation traces are stored in a tree where each node is associated with a state and its corresponding output value. Each edge is associated with an input value and a time step. The tester is based on the algorithm RRT for robotic trajectory planning [11]. First it creates a tree T the root of which is associated with an initial state. In each iteration, the tester determines a starting state **xstart** among the previously visited states stored in the tree T, and an admissible input value **u** and a time step h; it then supplies this information to the simulator. The simulator simulates the model from this starting state, under the corresponding input signal for a time h, this produces the new state **x** and the output **y**. The tester creates a new node in T and associates the new information with the node. It then connects the new node to the node of the state **xstart** by an edge labeled with the input value **u** and the time step h. The tester dynamics now can be rewritten as: $(\mathbf{xstart}, \mathbf{u}) = \mathcal{G}(T)$, $\mathbf{x} = \mathcal{F}(\mathbf{xstart}, \mathbf{u}, h)$, $\mathbf{y} = \mathcal{H}(\mathbf{x}, \mathbf{u}, h)$. In each iteration a starting state and an input value are chosen so that the new state they generate can improve the coverage [6]. Their computation is denoted by the function $\mathcal{G}$. The coverage is measured using the star discrepancy notion [4], which characterizes how well equidistributed a set of points is. Increasing the coverage enables exploring more behaviour patterns. To briefly illustrate the coverage metric, we assume that the state space is a box $B = [l_1, L_1] \times \ldots \times [l_n, L_n] \subset \mathbb{R}^n$. Let P be a set of N points inside B, which represent a set of visited states. Let us consider a sub-box J with the bottom-left corner coincides with that of B, and the other top-right corner lies inside B. The local discrepancy of the point set P with respect to the sub-box J is the difference between the ratio of volume and the ratio of number of points inside J, compared to the box B: $D(P,J) = \left|\frac{nb(P,J)}{N} - \frac{vol(J)}{vol(B)}\right|$, where $nb(P,J)$ is the number of points of P that are inside J, and $vol(J)$ is the volume of the box J. Now we move around the top-right vertex of the sub-box J inside B to obtain the set $\mathcal{J}$ of all such sub-boxes, and the star discrepancy of P with respect to the box B is defined as: $D^*(P,B) = sup_{J \in \mathcal{J}} D(P,J)$. The star discrepancy of P with respect to the box B satisfies $0 < D^*(P,B) \leq 1$. Intuitively, the star discrepancy is a measure for the irregularity of the point set P. A large value $D^*(P,B)$ means that the points in P are not much equidistributed over B, and the coverage of P is defined as: $Cov(P) = 1 - D^*(P,B)$. In the above definition, the sub-boxes are anchored at the bottom-left corner of the state space; it is possible to define other families of sub-boxes, such as those containing the centroid of the box B.

To improve the coverage, the function $\mathcal{G}$ determines, based on the current set of states in the tree T, a zone to explore. Such zones can be thought of large "holes" which still contain few visited states. Adding a new state in such a zone can improve the coverage; we call it *improvement zone*. Then, an existing state in T that is closest to this zone is determined to be the starting state for the next iteration. In an ideal case where the dynamics of the model, that is the function $\mathcal{F}$, is known, it is possible to find, by optimal control, an input function to drive the system from the starting state towards the zone. However, for Simulink models, the function $\mathcal{F}$ is generally not known, and we resort to choosing randomly an input value from the set of all admissible input values. Although this input value may not guarantee the new state to be closer to the improvement zone, the fact of choosing the starting state near that zone is crucial for coverage improvement and is proved to guarantee convergence towards the exact set of all reachable states for a class of models [6].

3 Adaptations for Simulink Models

State Reinitialization and Subspace coverage. We have assumed that all the state variables of the simulator are controllable by the tester. This is in general impossible, and the simulation tree T stores thus only the values of the controllable state variables (that is explicit state variables of Simulink blocks). The first adaptation concerns the ability of reinitializing all the state variables, in order to simulate the model from a given state. Here we are faced with a recompilation problem with Simulink. It is in general not easy to restore a previous state of the simulator. Our experience with the Simulink simulator indeed showed that reinitializing only controllable state variables (by reinitializing the initial conditions of the blocks) is not sufficient to fully restore the state of the simulator at a previous iteration. And saving "SimState" (which stores the hidden states among other information) could lead to a huge consumption of memory. We thus use the following solution. Since the simulator is deterministic when all the inputs are fixed and the model is not subject to noise, we retrieve the sequence of input values and time steps in the T that leads from the root to the state to restore, say $\mathbf{x}$. Then, we let the simulator restart the simulation from the initial state $\mathbf{x}_0$, under the retrieved input signal, as proposed in [9]. This guarantees the simulator to be in the state $\mathbf{x}$. On the other hand, it is possible, in the RRT setting, to sample a number of initial states and construct simultaneously from each one a simulation tree, which forms an RRT forest. We can vary the initial state of a Simulink model within its admissible set by changing the initial conditions of some Simulink blocks. This can be done by parametrizing these initial conditions which are then manipulated by the tester.

It is often of interest to focus on covering the domains of some critical variables, called *covered variables*, which have more influence on the satisfaction or violation of the property. This not only reduces computation complexity but also allows discovering interesting behaviours more efficiently. Indeed, the computation of the function $\mathcal{G}$ requires the star discrepancy estimation and geometric operations on point sets (such as the closest neighbours), which become expensive in high dimensions. To this end, we restrict these computations only on the subspace corresponding to the projection of the state space $\mathcal{X}$ on the covered variables.

Combining linear and branching traces. Covering the trajectories as much as possible allows discovering various behaviour patterns, but it may be costly in computation time and overly expansive in the exploration. It is of interest to be able to quickly reach a state which is known, from a-priori knowledge, to lead to critical behaviours. This is also useful for favouring long simulation traces, since state-space covering entails going back in time to start from a previously visited state. Therefore, we can optionally construct segments of long linear traces (in which the current state is also the next starting state) between the subtrees generated by branching the traces on different input signals. Since only one or several subtrees need to be explored further, in order to avoid simulating from the root of the tree, we could save the SimStates at the roots of these subtrees. As for the case of the HVAC model, this was used to delay branching the traces until a critical time point. This allowed us to detect property violations with much reduced computation time.

An interface between Simulink and the HTG tool have been developed [9] and recently enhanced to address the issues of the HVAC model. One role of the interface, implemented as a Matlab program, is to perform a procedure of locating the input signals to be controlled and the output signals to be observed by the tester. Additionally, there is an option to identify among $\mathbf{x}$ the covered variables. The interface is in charge of communication between the algorithms (that compute the function $\mathcal{G}$) of the HTG tool (implemented in C++) and the Simulink simulator (that computes the functions $\mathcal{F}$ and $\mathcal{H}$). The selection of the control inputs, initial conditions, and covered state variables can be guided by the testing strategies reflecting specific scenarios that the designer wants to test, as well as the property and the a-priori knowledge of the system provided by the designer. This will be illustrated by the dependency flow graph in the next section when we apply the proposed approach to the HVAC model.

4 Application to the HVAC Model

This case-study is modelled in Simulink with a supervisory controller used to optimally regulate HVAC components. This supervisory controller aims to enhance the functionality of interactive control strategies leading towards energy efficiency and a more user friendly environment. Verifying the supervisory controller performance is of a great value as it is the highest source of energy consumption in the building operation, and the most effective systems on user-comfort [16]. The main challenge in verifying these systems is their high complexity due to the high dependability of the variables. This dependability is introduced through the tight thermal coupling among all variables, which leads to a huge search space to verify the efficiency of the controller operations. In this section we demonstrate the above-described approach on this industrial HVAC model. By the usual simulation approach (such as on "corner cases" and a number of (randomly) chosen input signals and values of model parameters) no property violation was detected; however using our approach we were able to show the configurations which lead to property violations.

Figure 1 shows an industrial HVAC system modelled in Simulink. In this model, the heating and cooling controls are independent loops. The setpoints for the heating and cooling are optimized based on inputs from different HVAC zones [13, 14]. Formulated in our testing framework, the problem is to test whether under some variations in the environment (specified as external inputs), a given property is always satisfied. As shown in Fig. 1, the model contains four main components. The environment component models the external weather conditions (temperature, humidity, CO_2) and the building schedule. The control component models the control loops

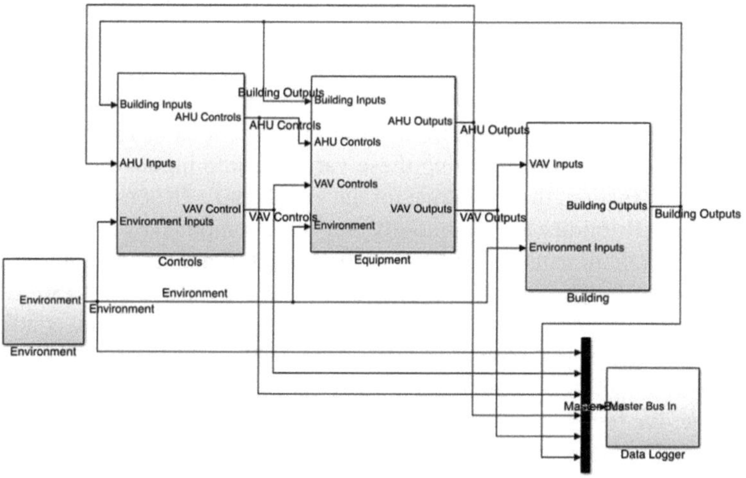

Fig. 1 HVAC model

for actuating AHU and each building zone VAV. The main objective of these control loops is to maintain the user comfort, while minimizing the energy consumption, where user comfort is respected by maintaining the indoor temperature, humidity and CO_2 at the predetermined setpoints. The equipment component models the AHU (Air Handling Unit) and the VAV (Variable Air Volume) actuation equipment, where the AHU is used to regulate the supplied air temperature to the building and the VAV regulates the air supplied temperature to each individual zone. The AHU uses two water coils to heat and cool, respectively, the supplied air by regulating the water flow inside each coil using water valve. The activation of each coil is determined based on the demand required. The building component models the building envelop, internal load, internal thermodynamics for temperature, humidity and CO_2. To show the complexity of this model, the command "model" of Simulink reported 270 (explicit) continuous state variables, 29 (explicit) discrete state variables. It additionally contains many Matlab functions and lookup tables.

Property. We are interested in a safety property: the heating and cooling should not be activated at the same time. That is, the outputs of the "Heating Control" and "Cooling Control" subsystems should not be greater than 0 at the same time, as the outputs of these blocks are the opening (in %) of the cooling and heating valves, respectively. Activating cooling and heating valves at the same time leads to a high energy waste.

Control inputs and Fault injection. The top-level environment component contains the external inputs to the system. Initially they are all set to constant. One of the testing strategies to limit the search space of the interdependent variables is to create dependency flow graphs to determine a set of input variables and internal variables which influence the variables involved in the property. As an illustration, the dependency flow graph for the Heating Coil Valve Control is shown in Fig. 2 (left).

The inputs of interest to vary (that is those with strong effects on the behaviour of the system) are: outer air temperature, humidity (OAT, OAH) and CO_2. The other inputs remain fixed. The variation ranges are: $OAH \in [50, 100]$, $OAT \in [10, 45]$, $CO_2 \in [600, 1500]$. It is also possible to study the system under variations of some internal variables, reflecting possible external perturbations or fault injection. This can be done directly by transforming these variables into inputs. In the experiment described in this section, this was done for the variables T_m (Room Air Temperature) and HR_m (Room Humidity Ratio). Initially they were computed by a Matlab function; we disconnected them from the outputs of the Matlab function and connect them to two new input ports. The variation ranges are: $T_m \in [15, 30]$, $HR_m \in [50, 100]$.

Initial state. We also varied the initial state, in particular the initial conditions of the integrator of the zone temperatures. We let them be in the interval $[19, 23]$, while in the original model, they were fixed at 22 (Celsius degrees).

Coverage. We focused on covering the states of the integrator block, which is an important continuous component modelling the zone temperatures.

We have performed a number of simulation runs, with different choices of input signals and covered variables, and obtained the following validation results.

Scenarios without detected property violation. The initial conditions of the integrator are fixed at 22. Figure 2 (right) shows the visited states in the projection on x_1 and x_2 which are covered state variables.

Scenarios with detected property violation. By allowing the initial conditions of the temperature integrators to be randomly chosen in [19, 23], and restricting the interval of T_m to be [25, 30], we could detect a time interval during which both valves are open. Three detected property violation scenarios are: (1) the initial zone temperatures in the integrator block are set to 20 (instead of 22 of the nominal regime), and $OAT \in [10, 45]$, (2) the initial zone temperatures in the integrator block are selected in [19, 22], and OAT is fixed at 22 as in the nominal regime, (3) a noise of 10 % is added to the variable T_m. Figure 3 depicts the temporal evolutions of the valve outputs for the first scenario.

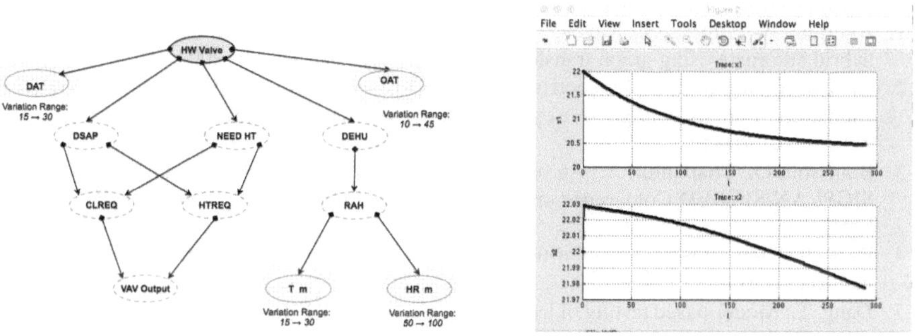

Fig. 2 *Left* Dependency flow graph for control input selection (Heating Coil Valve Control). *Right* No property violation detected. The figure depicts the temporal evolutions of two covered variables which are the states of two integrators. The thickness of the curves indicates the set of trajectories

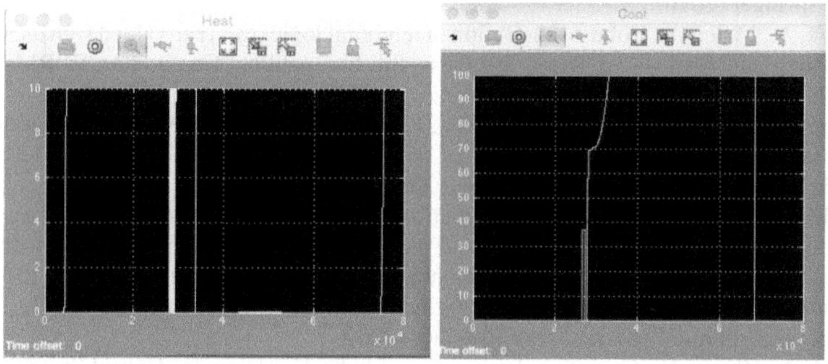

Fig. 3 Temporal evolutions of the two valve outputs. A property violation scenario: the initial zone temperatures in the integrator block are set to 20 (instead of 22 of the nominal regime), and $OAT \in [10, 45]$

5 Conclusion

In this paper, we extended the hybrid systems testing techniques to industrial Simulink models and demonstrated this result on an HVAC system. This result contributes an automatic semi-formal validation technique for cyber-physical systems, which are in general too complex to test manually. The experimental results are promising and we intend to pursue this work further by considering more complex properties (such as specified in Signal Temporal Logic [8], and using compositional testing to handle the complexity of industrial models.

References

1. Agrawal, A., Simon, G., Karsai, G.: Semantic translation of Simulink/Stateflow models to hybrid automata using graph transformations. ENTCS **109**, 43–56 (2004)
2. Arrieta, A., Sagardui, G., Etxeberria, L.: A model-based testing methodology for the systematic validation of highly configurable cyber-physical systems. In: VALID 2014, vol. 66–72. ARIA XPS Press (2014)
3. Bouissou, O., Chapoutot, A.: An operational semantics for Simulink's simulation engine. In: SIGPLAN/SIGBED Conference on Languages, Compilers and Tools for Embedded Systems (2012)
4. Beck, J., Chen, W.W.L.: Irregularities of distribution. Cambridge University Press, Acta Arithmetica (1997)
5. Dang, T.: Model-based testing of hybrid systems. In: Model-Based Testing for Embedded Systems, CRC Press (2011)
6. Dang, T., Nahhal, T.: Coverage-guided test generation for continuous and hybrid systems. Formal Methods Syst. Des. **34**(2), 183–213 (2009)
7. Donzé, A.: Breach, a toolbox for verification and parameter synthesis of hybrid systems. In: *Proceedings of International Conference on Computer Aided Verification*, CAV'10, pp. 167–170. Springer (2010)
8. Donzé, A., Maler, A.: Robust satisfaction of temporal logic over real-valued signals. In: Formal Modeling and Analysis of Timed Systems—8th International Conference, FORMATS 2010, LNCS 6246, pp. 92–106. Springer (2010)
9. Dreossi, T., Dang, T., Donzé, A., Kapinski, J., Jin, X., Deshmukh, J.V.: Efficient guiding strategies for testing of temporal properties of hybrid systems. In: NASA Formal Methods NFM 2015, LNCS 9058, pp. 127–142. Springer (2015)
10. Hoxha, B., Bach, H., Abbas, H., Dokhanchi, A., Kobayashi, Y., Fainekos, G.: Towards formal specification visualization for testing and monitoring of cyber-physical systems. In: International Workshop on Design and Implementation of Formal Tools and Systems (2014)
11. LaValle, S., Kuffner, J.: Rapidly-exploring random trees: Progress and prospects. In: Workshop on the Algorithmic Foundations of Robotics (2000)
12. Lee, E.A., Zheng, H.: Operational semantics of hybrid systems. In: Hybrid Systems: Computation and Control (HSCC), LNCS, vol. 3414, pp. 25–53. Springer (2005)
13. Mady, A.E.D., Provan, G.M., Boubekeur, M.: Towards integrated hybrid modelling and simulation platform for building automation systems; First models for a simple HVAC system. In: Information Technology & Telecommunication Conference (IT&T), pp. 191–199 (2009)
14. Mady, A.E.D., Provan, G.M., Ryan, C., Brown, K.N.: Stochastic model predictive controller for the integration of building use and temperature regulation. In: Conference of Association for the Advancement of Artificial Intelligence (AAAI), pp. 1371–1376 (2011)

15. Satpathy, M., Yeolekar, A., Ramesh, S.: Randomized directed testing (redirect) for Simulink/Stateflow models. In: Proceedings of the 8th ACM International Conference on Embedded Software, EMSOFT '08, pp. 217–226. ACM (2008)
16. Scenarios for a clean energy future: Interlaboratory working group on energy-effcient and clean-energy technologies (2000). NREL/TP-620-29379; ORNL/CON-476; LBNL-44029
17. Tripakis, S., Sofronis, C., Caspi, P., Curic, A.: Translating discrete-time Simulink to Lustre. ACM Trans. Embedded Comput. Syst. **4**(4), 779–818 (2005)

Modelling and Simulation of the Dynamics of Complex Socio-Cyber-Physical Systems and Large Scale Systems of Systems all Along Their Lifetime

Nguyen Thuy

Abstract This paper presents an innovative approach for the engineering of complex SCPSs and large scale SoSs. It is based on the modelling and co-simulation of the dynamic phenomena determined by the different disciplines involved in the engineering of such systems. This covers the complete system lifecycle, from prospective studies aiming at defining the nature and scope of a system to be developed, down to system operation and maintenance, retrofits and modification.

1 Introduction

To meet the needs and expectations of their users and customers, to satisfy ever more stringent safety, security and environmental regulations, to face with acute competition in open markets, industrial systems like power stations, aircrafts, vehicles must now be conceived as integrated *socio-cyber-physical systems* (SCPSs). Huge complexity, high societal expectations in terms of performance, safety, security and dependability, often stringent deadlines and budgets, very long lifetimes (up to many decades), sky-high costs and numerous and wide uncertainties: all these contribute to put pressure on the shoulders of system owners, designers and operators. Modelling and simulation all along system lifecycle could be of great help, but languages that do allow modelling all along lifetime are in a large part informal or semi-formal: they are not precise enough for extensive tool-supported simulation. Languages that are more suitable for simulation or computation need to be based on detailed design, and thus cannot be used in preliminary phases (prospective studies, system requirements specification, system architectural design).

Beyond individual SCPSs, which are designed, constructed and operated individually as integrated wholes, there are large scale, geographically distributed

N. Thuy (✉)
EDF R&D, 6, Quai Watier, 78400 Chatou, France
e-mail: n.thuy@edf.fr

G. Fanmuy et al. (eds.), *Complex Systems Design & Management*,
DOI 10.1007/978-3-319-49103-5_6

systems like national or continental power grids or railways systems. Such *systems of systems* (SoSs) are composed of loosely integrated constituent parts (often called *assets*) that may be complex SCPSs of their own. Determining which new assets need to be constructed and to which requirements, which existing assets needs to be retrofitted or upgraded or, on the contrary, should be phased out, is a key and existential issue for system owners like EDF, and is the domain of *prospective studies*. Such studies are often performed using models, but often with few connections with the engineering aspects.

Together with a number of other industrial organisations, tool providers and academic and scientific bodies, EDF R&D is developing an innovative approach for the engineering of complex SCPSs and large scale systems of systems (SoSs). This approach is based on the modelling and simulation of the dynamic phenomena affecting such systems, covering their complete lifecycle from prospective studies aiming at defining the nature and scope of a system to be developed, down to system operation and maintenance, retrofits and modification. The approach aims at the following objectives:

- Modelling of physical, cyber and human aspects.
- More generally, multi-aspect modelling.
- Massive simulation.
- Management of size and complexity.
- Systems reconfiguration.
- Validation of system behavioural requirements specification.
- Step-by-step design verification all along system development.
- Design optimisation.
- Diagnostics.
- Operation & maintenance optimisation.
- Reconciling innovation and safety and dependability.

2 Main Notions

The methodology is based in a large part on the FOrmal Requirements Modelling Language (FORM-L) that has been specified in the framework of the ITEA2 project MODRIO. Its scope is the formal modelling of **properties**, in particular of **requirements** and **assumptions**, in the form of **envelopes** of dynamic, time-dependent phenomena (see Fig. 1). Formal here means that the properties have rigorous syntax and semantics that can be interpreted by software tools to actively support a variety of systems engineering activities.

The modelling of a property addresses four main questions:

- WHAT constraints are to be satisfied? FORM-L constraints are either **Boolean conditions** or constraints on the number of **occurrences of an event**.

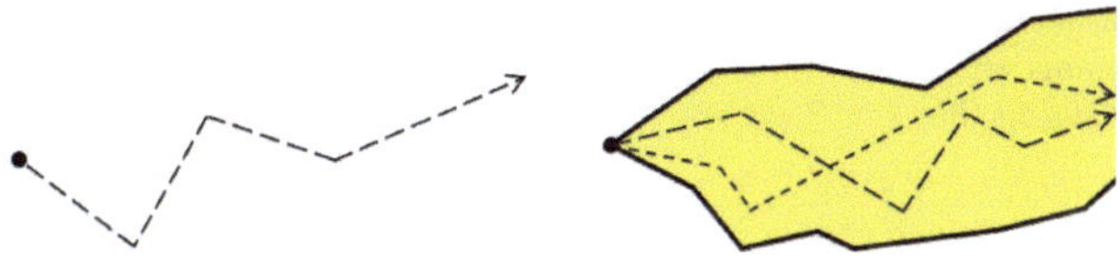

Fig. 1 Individual trajectories, versus envelopes allowing multiple trajectories

- WHEN are they to be satisfied? **Time locators** allow a precise specification of the time periods or of the instants where a constraint must be satisfied. The following simplified example specifies that when a command to open (*eOpenCmd*) is issued, a switch must reach the *open* state (an event) within half a second:

  after *eOpenCmd* ***within*** *0.5*s* ***check*** *open* ***becomes true****;*
- WHERE in the system are they to be satisfied? At early stages of the lifecycle, when the system architecture is yet unknown, the constituent parts of the system concerned by a property cannot be identified individually **in extension**. **Spatial locators** use the notions of set and set quantifiers (universal or existential) to specify the criteria that identify these parts **in intention**. The following example specifies that pumps not operating in emergency conditions should not cavitate (*pumps* being the set of all pumps in the system, and *emergency* and *cavitates* being attributes of a pump):

  forAll *p* ***in*** *pumps* ***suchThat not*** *p.emergency* ***check not*** *p.cavitates;*
- HOW WELL are they to be satisfied? FORM-L makes a distinction between desirable properties (which may be violated, as in real life systems failures are doomed to occur) and genuine requirements. A requirement generally requires the satisfaction of a desirable property under given fault-tolerance conditions, or puts a limit on the probability of not satisfying the property:

  check probability *(desirable.****violated becomes true****)* $< 10.^{-3}$*;*

Another objective of the FORM-L language is to integrate, on a **neutral ground**, the various pieces of information provided by the detailed models developed by the many engineering teams and disciplines that contribute to the system (see Fig. 2). For this purpose, FORM-L has the following concepts:

- **External** information and **bindings**. In most cases, a FORM-L model puts constraints on pieces of information that will be determined in more detail at later stages of the lifecycle or by specialised disciplinary models. Such pieces of information are represented in a FORM-L model by **variables**, **events** and **objects** that are declared as being **external** to the FORM-L model. When the model that generates a needed piece of information is available, then one can

Fig. 2 A neutral integration ground for the teams and disciplines working on the system

develop a binding that retrieves it and performs the necessary transformation to adapt it to what the FORM-L model expects, without having to modify the source model.

In the previous example regarding pump cavitation, the *cavitates* attribute is likely to be external. A physical behavioural model does not usually determine cavitation: rather, it calculates basic physical variables such as pressure and flow. The binding retrieves these physical variables, calculates the more functional attribute *cavitates*, and then provides it to the FORM-L model.

- **Contracts**. Whereas bindings can tie together models developed independently from one another, contracts between FORM-L models can be used to organise the cooperation of multiple teams. A FORM-L contract identifies the **parties** (i.e., the models) concerned, the **deliverables** (in the form of variables, events and objects) that each party must provide to the others, and the **properties** that apply to each deliverable. The party in charge of providing a deliverable views the associated properties as requirements. The other parties view the deliverable as being external, and the corresponding properties as assumptions (see Fig. 3). FORM-L also has **standard contracts** that can be applied to multiple sets of parties, and **contract extensions** that add deliverables and properties to an existing contract (in an inheritance-like manner).

3 Methodology

Modelling envelopes of physical aspects. Physical laws and phenomena are best described with multi-physics modelling languages such as Modelica. However, envelopes for physical quantities (which could represent assumptions regarding system environment, system requirements, or noise and measurements uncertainties), random events (such as deterioration of system components due to wear or

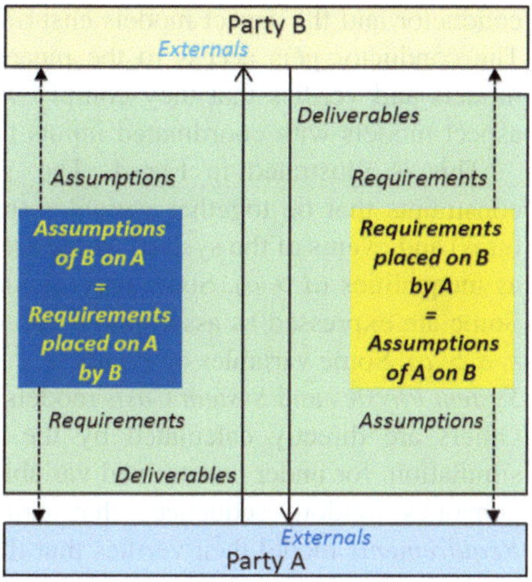

Fig. 3 FORM-L contracts

stress) and failure propagation (e.g., due to geographical proximity or physical connections) cannot be expressed in such languages, but can be represented in FORM-L.

Modelling envelopes of cyber aspects. Functional aspects (including the latent capabilities of smart equipment), data communication, networking and response times can be modelled using a variety of languages developed by the software and the embedded systems industries. Physical modelling languages and FORM-L can also cover these aspects. In addition, FORM-L can be used in preliminary phases to make sure that the requirements for the cyber parts of the system are consistent with those of the physical and human parts. It can also address failure propagation (e.g., due to functional dependencies or data communication) and common-cause failures (i.e., concurrent failure of multiple embedded subsystems due to the same design/software error).

Modelling envelopes of human aspects. FORM-L and envelope-based modelling can be used to represent the variability of the humans-system interactions, based on the results of psychological and sociological studies. In particular, the modelling can address human errors (one-off errors, or strategic error due to incorrect situation assessment). FORM-L can also be used to model human factor engineering requirements (e.g., "do not require human operators to make too many decisions and perform too many actions in a limited time", "give them the right information and time in case of an incident, so they can make a correct assessment and decide on the right strategy").

Multi-aspect modelling. There exist many languages for the modelling and simulation of the other aspects of a system: stochastic, economic, geometry, safety, etc. However, aspect models are often independent from one another. Therefore, their interactions cannot be evaluated using simulation. The proposed approach is to use a FORM-L model as an orchestra conductor, the bindings between the conductor and the aspect models ensuring information transfers in both directions. The conductor gets access to the pieces of information computed by the aspect models and verifies that they comply with the requirements. It also provides the aspect models with coordinated inputs that are consistent with the assumptions.

This is illustrated in Fig. 4. The *System Requirements* model specifies the constraints that tie together variables (representing dynamic, time-dependent features) and events of the system. Some are expressed as equalities ($d = b * c$), others as inequalities ($d > a$). Some may include temporal conditions (*before e*, *after e*). Some are expressed as assumptions ($5. < a < 10.$), others as requirements (*before e*, $d > a$). Some variables or events (*e*, *c*, *b*) are provided by specialised models (the *System Physics* and *System Costs* models) representing specific facets of the system. Others are directly calculated by the *System Requirements* model (*d*). During simulation, for under constrained variables and events, a Test Case Generation tool generates random sequences that comply with the assumptions. The *System Requirements* model then verifies that the requirements are satisfied. (This style of models organisation where physical models are directly bound to the system requirements model is in practice applicable only to very simple systems. A subsequent figure shows a more modular modelling approach suitable to complex systems.)

Massive simulation. Envelope-based modelling enables massive simulation: a tool such as StimuLus (from ArgoSim) can **automatically generate** any number of different sequences consistent with the specified assumptions, and **automatically check** whether for each sequence, the specified **requirements** are satisfied, violated, or not challenged at all. However, the space of possible situations the system can face is immense and impossible to cover exhaustively. To guide the tool towards sequences of particular interest, one can specify **generic test scenarios** in the form of additional FORM-L assumptions. Such a scenario is not a one-off

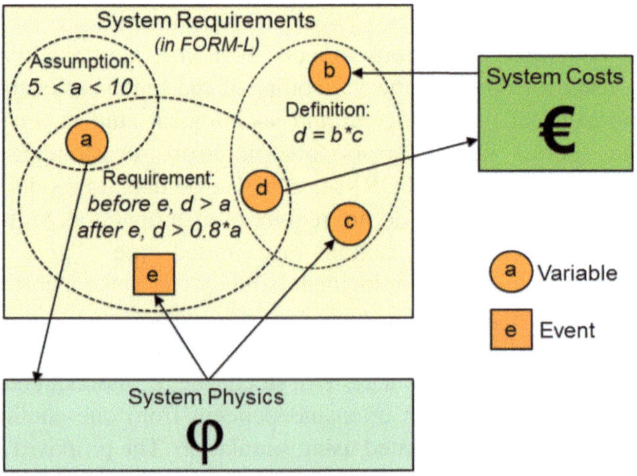

Fig. 4 A FORM-L model as an orchestra conductor

sequence: it is an envelope. Sequences can also be generated to satisfy **test coverage criteria** (to guide the tool in the systematic exploration of a model) and/or to **challenge** the requirements.

Management of size and complexity. Classical behavioural modelling is often made at a low level of detail, and thus is impractical for complex SCPSs and large scale SoSs. The proposed approach is first based on FORM-L **contracts** and **bindings** to support models **modularity** (where different system aspects or parts are represented by separate models), models **composition** (where multiple models are put together to more completely represent the system) and **abstraction** (where aspects, parts and/or details not relevant to a given study are modelled coarsely as envelopes, or left out altogether). Contracts offer a means for the **coordination** of multiple teams in **top-down** design phases: once the teams have agreed on a contract, each can work independently. If and when the contract needs to be altered or extended, this must be done with the agreement of all teams concerned. Bindings offer a means, in **bottom-up** design phases, to **reuse** something that already exists (e.g., commercial-off-the-shelf components such as smart equipment, existing environment entities, the current state of an SoS…) and that one does not wish to modify.

This is illustrated in Fig. 5, where, the *System* and its *Environment* are represented by two separate models. From the *System* standpoint, the assumptions are made regarding the *Environment*, whereas the requirements are applicable to the *System* itself. A contract formally specifies the mutual obligations of the two models. For example, variable *a* is provided by the *System Environment* model,

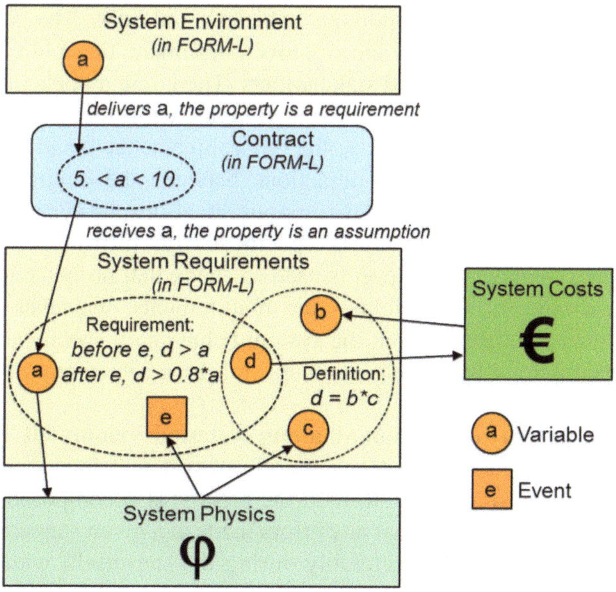

Fig. 5 Modular modelling

which guarantees that it satisfies *5. < a < 10*. The *System Requirements* model can now assume this to be true. The *System* and its *Environment* can now be modelled, studied and developed separately, provided that their models comply with contract. Contracts can also serve as an abstraction mechanism: when modelling and developing the *System*, for many system engineering activities, one just needs a simplified model of the *Environment*, one that satisfies the contract.

Validation of system behavioural specification. One of the first step of the engineering process is the development of system requirements specification, and of behavioural requirements in particular. For complex SCPSs, and for SCPSs that are to be constituent parts of an SoS, this is a difficult task involving many stakeholders with often incompatible **expectations**, many standards and **regulatory requirements** to be satisfied, and many different **situations** to be considered. Situations arise from the conditions provided by the numerous **entities** that constitute the **system environment**: other systems (which may be in various states, including failure states), human beings (who need to be informed, who may provide inputs, including incorrect ones, and who at each instant may set operational goals to the system, including contradictory goals and changes of mind), the physical environment (which may influence the normal operation of the system, and provide it with exceptional and hazardous conditions such as storms, earthquakes, flooding, extreme climactic events, …). Situations also arise from the system's own states (failure, maintenance, testing, operational goals at any given instant, …). Experience across all industrial sectors shows that even for safety and mission critical systems, errors in behavioural system specification are a significant source of failures, with sometimes unacceptable consequences. The proposed approach extends the one developed by MODRIO to address complex SCPSs, SCPSs in complex environments (e.g., autonomous vehicles), and SoSs. The principle is first to identify the entities constituting the system environment, the applicable standards and regulations, and the concerned stakeholders. Then, one models in FORM-L the assumptions made regarding each environment entity, together with the high level requirements placed on the system by the environment, the standards and regulations, and the stakeholders. The interactions between the environment and the system (viewed as a black box), and the various situations the system may face, are also modelled. The system behavioural specification is then modelled as a set of assumptions, and massive simulation is used to check that no behaviour consistent with the system specification violates the requirements, whatever the situation. FORM-L contracts are used so that the system behavioural specification is viewed as a set of assumptions by the environment, but as a set of requirements by the system itself.

Step-by-step design verification all along system development. Particularly in the case of complex systems, one does not go from system behavioural specification to detailed design in a single step. Rather, the process is decomposed in successive **refinement** steps. It is essential that any errors made at a given step are revealed and corrected as early as possible, preferably during the step itself, without having to wait until detailed design and accurate behavioural simulation, or worse, until operation. The proposed approach can be illustrated with the first design step, from system specification to overall architectural design, as shown in Fig. 6.

The *Preliminary System Design* represents one possible solution. It identifies the main components of the *System* (*X*, *Y* and *Z*), and provides a requirements model for each, of the same nature as the *System Requirements* model. The variables and events to be provided to the *System Requirements* model (*e* and *c*) are in fact provided by one of the components requirements models. Contracts between components specify their mutual obligations, much like the contract between the *System* and its *Environment*.

The *Preliminary System Design* can be verified by simulation as follows. Thanks to the contract between the *System Requirements* model and the components requirements models, the Test Case Generation tool views the components requirements as assumptions. It can generate any number of random sequences for the under constrained variables and events of the components requirements models, in compliance with these assumptions. The *System Requirements* model can then verify that the system requirements are satisfied.

For complex systems, purely random test case generation, even massive, is usually not very effective. There are several ways to guide the Test Case Generation tool to produce test cases of interest. One way is to develop generic scenario models that express additional assumptions. Another way is based on the notion of test coverage criteria.

When reaching detailed design, then one can also use accurate, specialised models such as Modelica. The proposed approach also provides a natural framework for Software-in-the-Loop and Hardware-in-the-Loop testing.

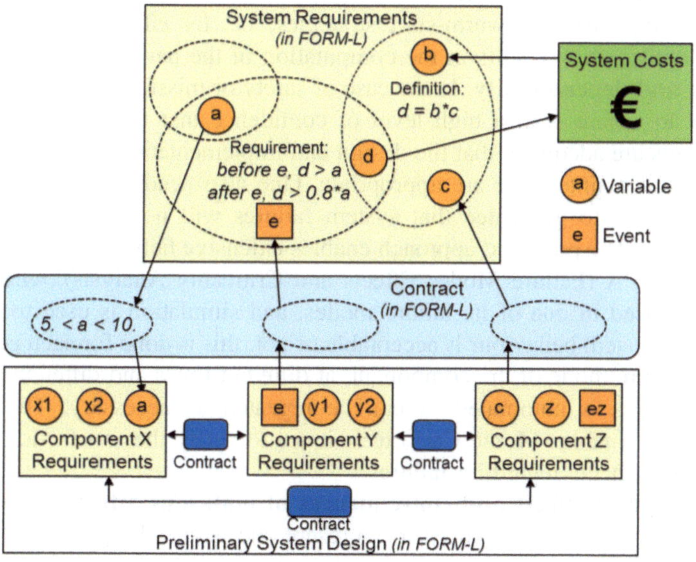

Fig. 6 Verification of a preliminary design

Design optimisation. For design optimisation, the proposed approach relies first on the ability to efficiently model, assess and verify multiple options at each step of the design process. With FORM-L's ability to model parametric designs and cost functions (including design costs, operation and maintenance costs, and revenues), massive simulation and/or analytical approaches may be used.

Systems reconfiguration. Reconfiguration occurs when SCPS subsystems, SoS assets, or environment entities are dynamically inserted or withdrawn. It is inherent to certain types of systems (e.g., traffic control systems) but could also be due to failures, maintenance, operational decisions, upgrades and modifications. With the notion of **dynamic set**, FORM-L can deal with reconfigurable systems. Extensions to Modelica are also being made to this end.

Diagnostics. To support diagnostics and operation in failure or accidental conditions, a simulator needs to be initialised to the same state as the real system. **State estimation** in failure condition is particularly thorny in the case of hybrid systems. The proposed approach ensures that design provides sufficient information (e.g., from enough sensors, or from embedded smart devices) to make a reliable state estimation.

Operation and maintenance optimisation. Optimizing the operation of a complex SCPS is a challenging task for which no general and robust method exists. The proposed approach is based on the notion of receding horizon, where an optimal control problem is iteratively solved based on the current state of the system, a physical model and a set of operational and physical constraints. To cope with the combinatorial and nonlinear aspects of the large reconfigurable systems, approximation techniques may be applied either to the system model or to the optimal control problem to get a numerically tractable formulation. The operation of CPSs in complex environments with tight real-time and safety constraints can be solved by implementing warm-start strategies, i.e. by efficiently exploiting the previous optimization results in the computation of the new optimal trajectories.

Safety and dependability. In the case of safety or mission-critical systems, it is necessary to ensure with a high level of confidence that the system behavioural requirements are adequate, that the design and implementation are correct, and that operation and maintenance are appropriate. One also needs to perform extensive failure analyses, to guarantee that system failures will not lead to unacceptable consequences. The proposed approach enables extensive failure analysis techniques such as FMECA (Failure Modes Effects and Criticality Analysis), where a component is placed in one of its failure modes, and simulation is used to determine whether the system behaviour is acceptable or not; this is done for each component, and each failure mode of the component, at different times and different situations. As this process is automated, it can be repeated as necessary. The proposed approach also enables Monte Carlo techniques to verify the satisfaction of probabilistic requirements regarding failures.

Assessment of effects and consequences of malicious attacks. The proposed modelling approach may be used to represent and assess the propagation and the consequences of successful attacks, so that one can identify where defences are most needed.

Reconciling innovation and safety. The proposed approach supports innovation by facilitating the assessment of alternative solutions at each step of the engineering process, but also by encouraging the explicit statement of the issues to be addressed: innovations have often resulted from a new way of stating an issue. At the same time, it support extensive, rigorous verification, including in failure conditions.

4 Conclusion

The approach being developed by EDF R&D is largely based on modelling and simulation all along system lifecycle. As a consequence, besides supporting many systems engineering activities, it also ensures that simulators are available very early to also support non-engineering activities, e.g., to help convince stakeholders and other decision makers. It also ensures that full scale simulators (for training or operation) are a very natural by-product reflecting the true design of the system. Lastly, whereas FORM-L directly answers the WHAT, WHEN, WHERE and HOW WELL, and the detailed models answer the HOW, by linking together the various viewpoints on the system and its environment and the various engineering phases, the proposed approach can answer the WHY. This is particularly important when retrofitting, upgrading or extending SCPSs and SoSs, as they tend to have very long lifetimes (some can even be considered as eternal) far exceeding the professional career of any individual.

Defining a Distributed Architecture for Smart Energy Aware Systems

Guillaume Habault, Jani Hursti and Jean-Marie Bonnin

Abstract In the past years, energy demand has increased and shifted especially towards electricity as the form of consuming energy. As the number of electric devices constantly grows and energy production must increasingly rely on renewable sources, this leads into noteworthy disparity between electricity production and consumption. This paper describes the results of a joint work between partners in ITEA2 12004 Smart Energy Aware Systems (SEAS) project, which aims towards providing the ICT tools and systems in order to help energy actors better manage and optimize energy consumption, production and storage. This paper presents and studies the innovative IT architecture proposed during this project, SEAS Reference Architecture Model (S-RAM). This architecture relies on four distributed services that enable to interconnect any energy actors and give them the opportunity to provide new energy services. The benefits of S-RAM have been studied on a specific use case, which aims to provide a service for estimating local photovoltaic production. It particularly helps energy management systems better plan electric consumption.

1 Introduction

There are several notable trends taking place in the energy market today. These include for instance the shift away from oil as an energy source especially in transportation, the steadily increasing number of various electric appliances and

G. Habault (✉) · J.-M. Bonnin
Institut Mines-Télécom/Télécom Bretagne, 2 rue de la Chataigneraie, 35510 Cesson-Sévigné, France
e-mail: guillaume.habault@telecom-bretagne.eu

J.-M. Bonnin
e-mail: jm.bonnin@telecom-bretagne.eu

J. Hursti
ASEMA ElectronicsOy, Betonimiehenjuja 3, 02150 Espoo, Finland
e-mail: jani@asema.com

G. Fanmuy et al. (eds.), *Complex Systems Design & Management*,
DOI 10.1007/978-3-319-49103-5_7

devices, and the decreasing prices of distributes energy production technologies such as solar panels.

The majority of the changes lead towards a system where central energy production facilities—dams, nuclear power plants, etc.—must co-exist with a myriad of smaller, less reliable systems in the same network while at the same time energy demand for them will show a significantly higher fluctuation.

Because proper operation of the electrical network is based on the balance between production and consumption, this poses a great challenge for the management of the network. To properly cope with the problem, new IT systems are needed for energy actors to interconnect and better manage energy use. The SEAS Reference Architecture Model is an architectural model that draws from the best practices of Internet technologies, especially the Internet of Things (IoT) in order to define solutions to the basic problems such a balancing system must face.

For the system to be properly balanced, real-time and predictive measurement along with control capabilities are needed in a widespread management system. This necessarily involves handling the issue of controlling a large volume of distributed consumption and production points which can simultaneously act as an energy producer and/or consumer hence the term "prosumer". While remotely controlling and coordinating the electrical loads of homes, office buildings and industrial premises has been possible for decades already, such controls are not yet widely enough adopted to confront the challenges of new electrical networks.

A key reason for this is that such adoption is still too expensive in high volume. The process is costly because it involves

- Finding each party that acts as the gatekeeper to some resource (the "finding problem")
- Receiving the security clearance for accessing the resource (the "access problem")
- Learning the details of the mostly proprietary access method (the "compatibility problem")
- Implementing the technical compatibility to each remote system (the "implementation problem")
- Managing the monetary compensation for the access and ensuring compliance to commitments (the "compensation problem")

To solve these problems, technology is needed to offer automated access to each measurement point and load to lower the cost of control for the service provider. The S-RAM architecture has been designed as such a solution and addresses each of the core problems listed above.

The core of S-RAM is a set of directory and security services, the Core Services that in turn coordinate the efforts of various distributed Group Managers and End Nodes. The traffic between these nodes is modeled according to the principles of semantic data to remove the ambiguity of data interpretation and offer a common, high-level protocol language for this purpose.

In the S-RAM architecture, each node registers itself to the Registry Service, which maintains a map of management relationships. Access can then be established by a new party even on ad hoc basis by finding the proper resource from the registry and negotiating access rights to it with the help of trusted third party authentication. Once a control action is performed, this is recorded in a Transaction Service, which holds an audit log and enables paying back a compensation for the control.

The rest of the paper is structured as follows. First, existing architectures and solutions are overviewed and compare to SEAS project requirements. In Sect. 3, we present the terminology and a general description of the architectural model defines for SEAS project. The main principles and their advantages are briefly presented in the Sect. 4. Before concluding this paper in Sect. 6, we present some of the results obtained with a proof-of-concept implementation of this model based on a PV production estimation service.

2 Related Works

The partners of SEAS project have defined over 100 Use Cases (UCs) that could demonstrate the benefit and possible applications of the architecture. It has consequently been of crucial importance to determine the requirements of those UCs—which can range from network to functionalities—before overviewing existing architecture and solutions.

The study of the SEAS UCs along with the structure of current energy networks and the forecasted development, stressed the fact that SEAS project required an innovative ICT architecture to address all its needs. While different types and levels of management are envisioned in SEAS project—from Area management (Home, Building, Microgrid, Regional) to Device control management—all of them still aim at coordinating and optimizing energy production and consumption. The core of such management is the collection of data from different types of nodes—constrained or not—and the analysis of collected information. Various actors also need to send energy solicitation or energy demand in order to make energy optimization possible. As a consequence, ICT tools and architecture used to support SEAS project should be (a) scalable, enabling interconnection of billions of energy nodes along with energy systems; (b) dynamic, making it possible to adapt and change the system as requirements evolve; (c) automated, facilitating the integration of systems and an exponentially growing number of nodes; and (d) secure, ensuring access control and data privacy along with providing secure communications.

Several architectures exist that aim at solving Internet-of-Thing (IoT) issues, to cite but a few [1–3]. All of them make it possible to connect different nodes and systems, to retrieve data from endpoints, and to control the nodes. None of them however provides the type of high volume mapping and search capabilities that energy network operators are looking for to cope with the dynamism and automation requirements of modern grids.

Several other systems exist to manage the energy Demand-Response (DR) required in energy network such as [4, 5]. However, these systems are costly to implement and for instance require one interface per service to interconnect. Consequently, these solutions lack the adaptability needed by SEAS.

Finally, several management systems are available to collect and analyze data among which are [6–8]. These management systems cannot be used as a centralized architecture for SEAS project, as it will not provide the required scalability and availability. However, these systems could be integrated to the chosen solutions.

As none of the known solutions address all the requirements for SEAS project, it was decided to define a new architecture model. The model we propose, called SEAS Reference Architecture Model (S-RAM), can cope with all the unveiled requirements. In addition, this model is based on some known best practices and reuses some of the concept and principles from [1] and adds new ones.

3 SEAS Reference Architecture Model (S-RAM)

3.1 Terminology

As mentioned previously, the SEAS project has defined several UCs that can be used to demonstrate the project benefits. These UCs apply on different domains, from domestic and building management to micro management but also Electric Vehicle (EV) charging, etc. To be able to map the architecture to any UC, a common terminology was first applied to all the UCs. As shown in Table 1, there are two main categories of entities, which can play different roles. This list in the table is not exhaustive but gives an overview of possible roles.

3.2 General Description

The objective of S-RAM is to enable secure measurement and control of energy loads in a manner that is sufficiently well-defined but not so rigid it would hinder innovation. In order to satisfy this purpose, the architecture should provide the tools for:

1. An energy manager to retrieve energy production, consumption and storage information– referred as ***energy information*** in the rest of the paper; and
2. Other actors to provide systems, mechanisms or services that use the energy information for energy management and control.

S-RAM uses four core verbs to model the actions of parties: give, take, keep and alter. In the energy domain, these mean for instance producing energy, consuming

Table 1 List of SEAS communication entities

Type	Name	Roles	Example of matching
SEAS field entity (SFE)	End user (EU)	Interact with other entities to plan and control energy management	Resident, EV driver
	End node (EN)	Entity consuming, storing or producing energy or the node monitoring such entity	EV, Production unit
	Non-SEAS EN (NSEN)	An EN that does not support SEAS communication mechanisms	Home appliance
	Node controller (NC)	Entity capable to store data from different NSEN or EN. It might aggregate, forward and if required translate data into SEAS model	EVSE
	SEAS group (SG)	A group of SFEs. It can for instance be mapped to geographical area	Building, microgrid
SEAS core entity (SCE)	Group manager (GM)	Entity managing one or several SGs. It has the capabilities to store and analyze data collected within an SG	Energy management Systems
	Energy distribution operator (EDO)	Entity distributing energy to an SG	Distribution system operator
	Energy market operator (EMO)	Entity managing energy market	Day-ahead prices
	Energy service provider (ESP)	Entity providing an energy service to other SCEs	

energy, storing energy to a battery and alter the states of consumers to for instance sell flexibility in consumption, i.e. negative energy.

The four services listed in Table 2, called SEAS Core Services (SCSs), are necessary to fulfill these objectives. These SCSs make the coordination possible while ensuring the architecture to be automated, adaptable and secured.

Table 2 List of SEAS Core Services

Name	Description
Registration service	Enables any trusted SCE to register itself and its capabilities (Management of SG or Provider of an Energy Service) for others to find them. It also allows to subscribe to other services
Ontology service	Links data to the standardized vocabulary, which enables any SCE to interpret received messages making it interoperable
Transaction service	Connects SCEs to banks and payment systems and acts as a trusted notary to enable compensation for participation in energy management
Security service	Enables the authentication of any SCE willing to participate in the ecosystem and to communicate control preferences, thus helping access control

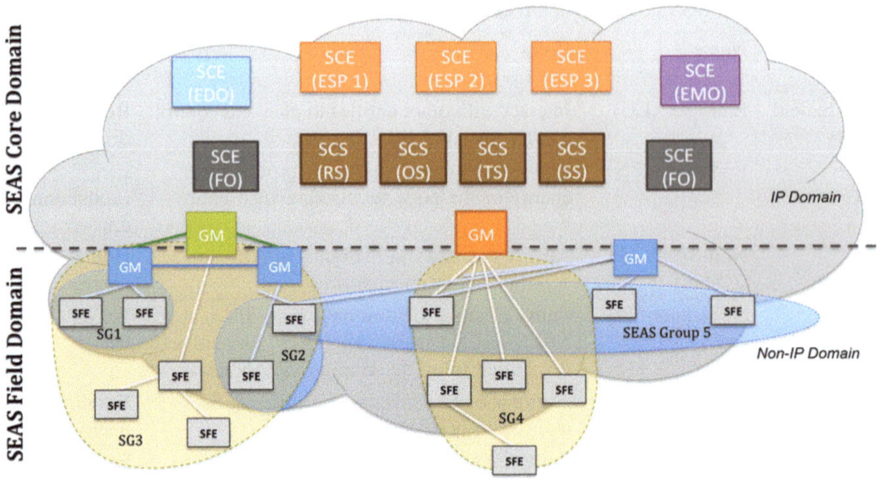

Fig. 1 Illustration of SEAS reference architecture model

S-RAM also depends on basic principles and a communication methodology in order to make the interconnection possible while making it scalable as presented in the following section. As depicted in Fig. 1—an illustration of the architecture—S-RAM is divided in two domains: SEAS Core Domain (SCD) and SEAS Field Domain (SFD).

SCD is composed of the four SCSs and several SCEs. SCD relies on Internet Protocol (IP) to interconnect all these entities and services. The SCSs function as trusted parties for finding and identifying parties, once done the SCEs communicate on a peer-to-peer basis.

SFD is composed of several SEAS Groups (SGs) each of them managed by a Group Manager (GM). AGM collects energy information from all SFEs belonging to its group, stores it and optionally processes it. In order to efficiently make energy plans and to be able to reply in the best way to any grid solicitation, GM should provide at least the following SG information:

1. Total energy consumption of the group and if any, total production and storage level;
2. (Optional) Production and/or consumption estimation;
3. (Optional) Load shedding capabilities.

Any of this three information can help an SCE make more relevant solicitation to a SG as regard to its energy consumption—i.e. a distribution operator demanding a SG to supply its group using its own production units.

3.3 Example with UC: Estimation of Local Photovoltaic Production

In this scenario, we consider the case of houses with photovoltaic electric production capabilities. These houses could help the grid flatten its distribution on peak hours by consuming their own production (directly or stored) for given periods on grid demand. It is assumed that the house has a House Energy Management System (HEMS) in place. However, as is typical, the HEMS faces the problem of ensuring that today's production can be sufficient to support the house consumption for a given period of time. To be able to respond to grid demand intelligently, an HEMS needs to be able to estimate the house consumption and production. In this UC, we focus on estimation of PV production as illustrated in Fig. 2.

In this scenario, three entities are required:

(a) An HEMS, to monitor and store PV production measurements;
(b) A Weather Forecast Service Provider (W-SP), to provide a cloudiness percentage service for a given geographical area;
(c) An Estimation Service Provider (E-SP), to estimate future production based on both future cloudiness percentage and historical production measurements and cloudiness forecast.

According to S-RAM terminology, the household is a SEAS Group (SG), the HEMS is a Group Manager (GM) and collects data from equipment and nodes within the considered SG. W-SP and E-SP are SEAS Core Entities (SCEs) that provide dedicated services—which are registered on the SEAS Registration Service (RS)—to other SCEs. As a consequence, the HEMS can search for an estimation service and the ESP for a cloudiness percentage forecast service for the applicable location from the S-RAM registry. When found, they can directly contact them in order to have access to the desired service—they might also use SEAS Transaction Service if required, e.g. to pay for the service.

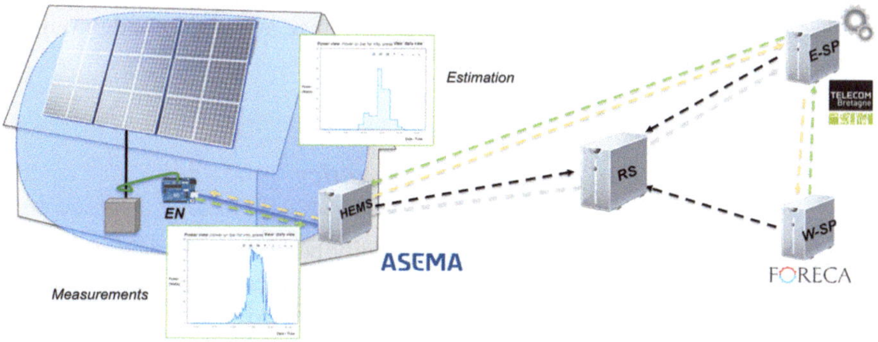

Fig. 2 Illustration of PV production estimation use case

The HEMS will then provide historical PV production measurements and its location to the E-SP. The E-SP will retrieve cloudiness percentage forecast from W-SP for this location. With these collected data, E-SP will compute future PV production and send it back to the HEMS.

4 S-RAM Principles

In this section, we will present the principles on which S-RAM rest upon. We believe that such architecture should rely on basic principles of availability, automation, scalability and security.

4.1 Distributed SEAS Core Services

The separation in two domains in S-RAM has several advantages. The first advantage is the separation of the constrained world (Field Domain) from the non-constrained one (Core Domain). The Field Domain may contain various constrained IoT devices that are deployed to monitor or control our environment—especially around energy. These constrained devices are not supposed to be used as such on Internet as their current protocols are not adapted to manage such an amount of devices. Gateways are used to regroup and properly link them to Internet.

The second advantage is that S-RAM allows the use of different types of communication protocols to be used within domains. The SCDs mainly use HTTPS and MQTT for communication. The SCEs may use a mixture of HTTP, MQTT and CoAP. The SFEs on the other hand might not be capable to use such protocols, not even IP. To facilitate this in the architecture, the GM has the responsibility of linking its SG to the SCD and its offered services and handling the possible protocol translations in between.

As a result, any SCE and especially the four SCSs will be available from any IP-capable node irrespective of their location. The SCS have been designed so that they can be distributed over several servers to maximize availability and throughput without the risk of Single Points of Failure.

4.2 Communication Relying on Semantics

S-RAM enables the application level compatibility between SCEs with a common application protocol that removes any ambiguity when interpreting data. A semantic data model has been designed for the messages of S-RAM. The S-RAM Ontology Service is used to distribute the model and a technology is being developed to make various nodes automatically pick up and adapt to the changes in the model. As a result, new services could be added all along the life of S-RAM without having to modify already existing nodes.

4.3 Different Type of Communication Modes

As aforementioned, communications in SCD principally rely on IP/HTTP(S). S-RAM defines their two types of communication mode, direct or hierarchical. The former one is quite simple, any SCE can communicate with another found SCEs (its services or any SFE belonging to a managed group, if any). The latter communication mode is used when SGs work in a hierarchical way. In fact, an SG might belong to another SG—e.g. to map the structure of a city, a house belongs to a sub-district that belong to a district. In such configuration, corresponding inter-GMs communication can be hierarchical. A higher-level GM can add the overall energy information of a sub-group in its own energy management mechanism. As a consequence, GM at a higher level of hierarchy does not need to have detailed information of sub-groups. It helps keep information privacy of sub-group inside the group. However, if a higher-level GM requires to adapt the total group consumption or production, it only needs to send generic energy solicitation objectives– as the grid would—to lower-level GM, which then handles the low level I/O decisions.

As a consequence, this domain separation helps S-RAM be more scalable. It also helps maintaining data privacy within a given group, as visibility can always be limited to a certain level of hierarchy.

4.4 Secured Exchange and Architecture

As the S-RAM architecture defines peer-to-peer communication, assisted by a set of SCSs, the security of the S-RAM architecture relies on the peers following the practices recommended by the architecture. The SCSs offer the infrastructure that allows these practices to be followed in a distributed network.

Data confidentiality and protection against tampering is built in S-RAM on standard Internet protocols HTTPS and DTLS. The S-RAM Security Service acts as a Certificate Authority that can issue certificates required by those protocols. It supports the SCEP for high volume deployment.

Authorization and access control should be handled by the peers. The certificates from the Security Service are used for identification. The Security Service also supports dissemination of access control rules through a trusted party.

Availability of the SCSs is designed in using session less messaging and load balancing between redundant entities. Ensuring the availability of end nodes and other services is left as the responsibility of the said parties.

Non-repudiation is assisted by the S-RAM Transaction Service. Any transaction needing non-repudiation can be notarized by the service.

5 Application for the Estimation of Photovoltaic Production

The UC used to illustrate S-RAM in Sect. 3.3 has been implemented to test its feasibility. The following presents both this UC set up and obtained results.

5.1 Implementation

In this UC implementation, we deploy a 1 × 0.6 m PV panel. An Arduino mini-computer acts as an End Node (EN) and monitors the electric production of this PV panel. Asema IoT Central software is used as the Group Manager (GM) Every 30 s the PV sends a CoAP POST request to the GM with a binary payload containing the panel production reading. The GM then applies templating technology to transform the binary on a semantic message (Turtle) that is interpreted and the data is stored.

Télécom Bretagne has developed a PV production estimation service, called TB-PVEstimation, for this scenario. In order to estimate future production, this service requires the following:

1. The geographical location of the considered PV panel;
2. The cloudiness percentage forecast for this location;
3. The electric production of the panel.

Foreca Ltd provides digital weather data service as an SCE and especially the hourly cloudiness forecast. Each day, the TB-PVEstimation collects and stores both hourly-based cloudiness percentage forecast from this SCE and hourly-based PV production from the GM. Historical data are then used with a machine-learning algorithm to compute a PV production estimation, which are then sent and visualized.

5.2 TB-PVEstimation Algorithm

The estimation algorithm developed for this implementation uses machine-learning python algorithm. It computes, for almost each hour of the day, a 1st degree polynomial function of PV production versus cloudiness percentage forecast. Therefore, for a given SG and a given season, TB-PVEstimation service computes less than 24 functions, depending on sunrise and sunset hours. All these functions

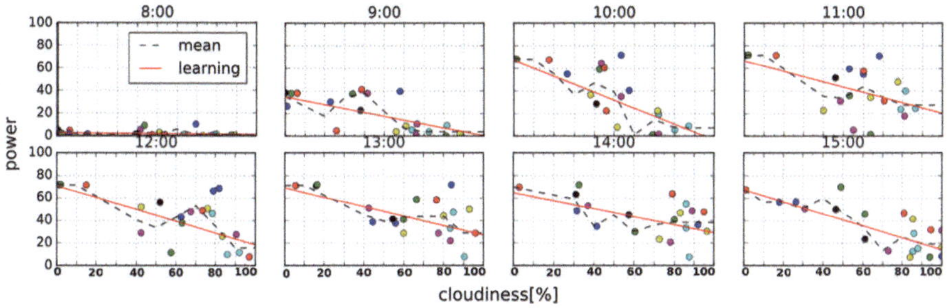

Fig. 3 Machine-learning versus average estimation functions

are re-evaluated every day when receiving latest hourly PV production measurements for a given SG.

Figure 3 represents some of these functions for this implementation. The color dots illustrate all the production measurements retrieved for given cloudiness percentage depending on the hour of the day. Dashed lines represent the average production based on collected measurements. Solid lines show the 1st degree estimation function. It is to note that the aim of this paper is not to compare our estimation algorithm to existing ones but to present how S-RAM can help Service Provider offers new energy services to Energy Management Systems.

5.3 *Results*

SEAS Registration Services enables Asema IoT Central to search for an estimation service and contact the selected one. As a result, the end user receives an estimation for the coming production of a PV panel.

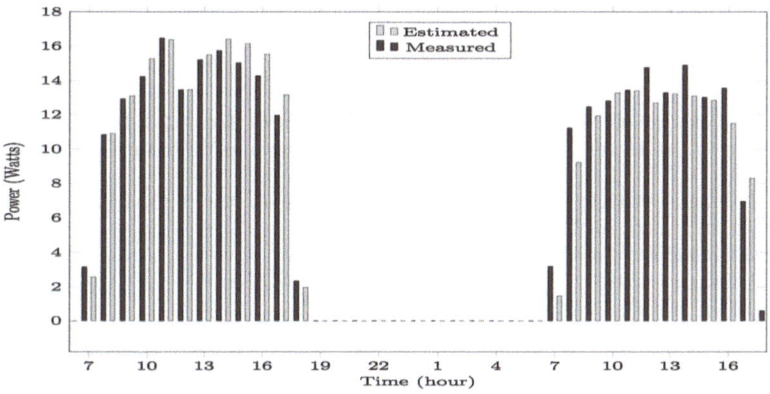

Fig. 4 Measured versus estimated production

Figure 4 presents a comparison of hourly PV production estimation with the actual measured production. It shows that the estimation service, after having collected two weeks of measured production, gives fairly good results and an average of 2 % of error for this panel. This error tends to decrease with time thanks to machine learning.

6 Conclusion and Perspectives

In this paper, we presented the SEAS Reference Architecture Model (S-RAM). This architecture concept aims to provide the ICT tools to interconnect energy actors in order to better manage, coordinate and optimize energy consumption, production and storage. We showed the usefulness of such architecture on a PV production estimation scenario. In this scenario, a PV panel owner can retrieve day-ahead production estimation for a panel and use the estimation production information with the help of an Energy Management System (EMS) to better plan energy needs and to inform the electrical grid on future needs and capabilities.

The S-RAM Registration Services enable any Service Provider to deploy new and innovative services for any EMS and make them automatically available for a large mass of users. Semantic information, supported by an Ontology Service, helps the EMS automatically understand and interpret information received from an end node.

We plan to extend the implementation of S-RAM and further test it with other scenarios, as well as realizing a safety analysis. It should end up demonstrating all the benefits offered by this architecture—adaptable, secure, automated and scalable. Several businesses and possibilities can be envisioned on top of S-RAM.

Acknowledgment This work has been partially funded by the ITEA2 12004 Smart Energy Aware Systems (SEAS) project—http://the-smart-energy.com.

References

1. oneM2M: Functional Architecture. Technical Specification, OneM2M. http://www.onem2m.org/images/files/deliverables/TS-0001-Functional_Architecture-V1_6_1.pdf (2015). Accessed Jan 2015
2. IoT-A: Final Architecture Reference Model. Project Deliverable, IoT-A. http://www.iot-a.eu/public (2013). July 2013
3. Industrial Internet Consortium: The Industrial Internet Reference Architecture. Technical report, IIC, June 2015
4. OpenADR: http://www.openadr.org/
5. TNO: Energy Flexibility Platform and Interface. White paper, TNO, June 2015
6. Sofia2: http://sofia2.com/
7. Fi-Ware: https://www.fiware.org/
8. Asema IoT-Central: https://www.asema.com/en/

Incremental Modeling Methodology of Railway System Specifications

Melissa Issad, Leila Kloul and Antoine Rauzy

Abstract Specification of complex systems is a set of large documents written in natural language. Due to their complexity, they are often hard to understand and even harder to maintain. We designed the domain specific language *ScOLa* (Scenario Oriented Language) to model the architecture and behavior of systems using a set of formalized concepts in order to support the dialog between experts. In this article, we present a reverse engineering methodology to formalize complex system specifications using scenarios. It starts from an informal description of the system and results in a hierarchical view of the system description. This article aims both at introducing *ScOLa* and at presenting its application on the railway systems.

Keywords *ScOLa* · Systems engineering · Formal specification · Modeling language · CBTC

1 Introduction and Motivations

A Communication Based Train Control (CBTC) system is an automation solution for railways. The Trainguard Mass Transit (TGMT) is the CBTC solution of Siemens that is the basis of our study. It equips driverless trains and is therefore responsible

M. Issad (✉)
Laboratoire Genie Industriel, CentraleSupélec,
Grande Voie des Vignes, 92290 Chatenay Malabry, France
e-mail: melissa.issad@siemens.com

M. Issad
Siemens SAS, 150 avenue de la République, 92320 Châtillon, France

L. Kloul
DAVID, University of Versailles, 45 Avenue des États-Unis, 78000 Versailles, France
e-mail: leila.kloul@uvsq.fr

A. Rauzy
IPK, Norwegian University of Science and Technology,
S. P. Andersens veg 5, 7491 Trondheim, Norway
e-mail: antoine.rauzy@ntnu.no

G. Fanmuy et al. (eds.), *Complex Systems Design & Management*,
DOI 10.1007/978-3-319-49103-5_8

for all the train functions. Hence, it consists of several sub-systems and functions to maintain train movement.

Design and development of such systems is based on a system specification. It represents a series of documents of up to a thousand pages each, that describes as explicitly and precisely as possible the system. These documents, produced by the system engineering teams, are made up of: a high-level system requirements specification, a system architecture specification, a performance specification, a glossary, and an interface specification. Written in a natural language, these documents describe the constituents of the system as well as its behavior in its different phases. They are used by software developers as a basis for their work, by validation teams to generate test cases, and finally by safety analysts to retrieve potential failures. The high-level system requirements specification consists of a set of requirements of two types: definitions and proses. Definition requirements are functions definitions and actions, proses are explanations on the context of the definitions. The system architecture specification describes informally the top-level components and top level functions. It includes elements about the environment of the system, redundancy of components and connections between components. Top-level functions are depicted using an informal description followed by a set of high level requirements. The performance specification provides additional requirements and calculations, for example regarding timing constraints. The glossary explains most technical terms and abbreviations used in the other documents. Finally, the interface specification explicits the data exchanged between components. In addition to the previous documents, a document called "functional specification operational scenarios" provides a number of operational scenarios of the system. Each scenario is introduced by a short description, followed by the initial conditions of the used components. Then, the scenario is described in details. The scenarios refer to requirements, components and communication channels and interfaces.

However, the use of natural language leads to several problems. First, it may cause ambiguities; words may have several meanings according to the culture and background of system engineers. Second, validation teams need to provide test cases at the system and sub-system levels. This requires two things: the functions descriptions have to indicate which sub-systems are involved, and there must be a synchronization between functions descriptions and system architecture. But, there are several cases where sub-systems are omitted from functions descriptions. Also, some descriptions depict sub-systems from different levels of the system architecture. Third, the functional decomposition is not optimal, most of the functions are not self-supporting. Hence, a single function description spreads in several functions. Finally, the system architecture, as it is depicted in specifications, prevents the full retrieval of the actual architecture of the system from these documents only, they are listed instead of being decomposed.

To tackle these issues, a potential solution consists in switching, at least partly, to model-based system engineering. Models provide an ideal vehicle for complex systems representation and abstraction. A model is a representation of a complex system using general rules and concepts. The objective in complex systems modeling is to find a suitable set of concepts to capture, in a single model, the architecture and behavioral parts of a system specification. Both of these parts must be synchronized. Moreover, the model must remain accessible for engineers to use it as a communication support. However, the introduction of models is quite difficult. First, because system experts may not be familiar with formalisms and modeling languages. Second, ambiguity and incompleteness of system specifications and also the functions descriptions that are spread in several documents prevent from a direct translation into a model.

There exist in literature and practice, several methods and languages addressing the identified issues. Some of them are formal and others semi-formal. The most remarkable use of formal methods into the specification of railway systems is the B-method based on the B language [1]. It consists in an incremental modeling of the system with proof and validation objectives. However, a prerequisite for such an approach is the completeness and non-ambiguity in the system description. It also requires an expertise in the B language which is rare and costly. Semi-formal methods rely on graphical notations that make it possible to represent different aspects of the system, by means of specialized views. These graphical notations aim mainly at being a communication support between stakeholders. SysML [2] is probably the most popular of these notations. However, it is not well suited for the reverse engineering of existing system specifications. Technical concepts in the documents have no direct representation. Some interpretation work is always necessary, which is both tedious and a source of ambiguity. Using different views turns out also to be quite problematic. It is actually difficult to warranty the coherence between the views and to ensure the completeness of the model as a whole. Moreover, graphical constructs are difficult to understand by non-specialists and could not achieve fully the description of systems. SysML provides the possibility to define the so-called profiles [2], that is to specialize SysML for particular needs. However, this approach is moderately convincing; it looses somehow the generality of the representation without really alleviating significantly the interpretation/comprehension and model validation work.

In this article, we present an incremental methodology to formalize the representation of the architecture and behavior of complex railway systems starting from an informal system specification written in a natural language. The methodology is supported by *ScOLa* [3], a scenario oriented modeling language dedicated to the analysis and formalization system specifications. Our attempts to use SysML (or any other existing notation as BPMN [4] or statecharts [5]) were not successful. We were spending more time in casting the concepts we need into the notation than to elaborate the concepts themselves. The objective of *ScOLa* is to be a compromise between two worlds: a graphical notation to represent a system specification and a set of formally defined concepts to elicit a model of the system. The contribution of this article is therefore threefold. First, it presents *ScOLa*, its ability to represent the architecture and behavior of a system specification, by means of examples. Second, it presents the reverse engineering methodology we applied to translate incomplete, ambiguous textual specifications of a full scale industrial railway system into more formal ones. Third, it shows the interest of the Domain Specific Language approach for modeling purposes. Although *ScOLa* was primarily designed to describe railway systems, we believe that it is suitable to a broader range of applications.

The remainder of this article is organized as follows. Section 2 presents the running example we work on. Section 3 presents the *ScOLa* language with both its textual and graphical representations. Section 4 defines an incremental methodology to define a *ScOLa* architecture and scenario model. Finally, Sect. 5 discusses related works.

2 Running Example

Our work consists in the formalization of the system architecture and operational scenarios of the railway TGMT CBTC system of Siemens. It is a train control system for metros, light rail systems and commuter trains. The TGMT system performs its missions through interactions between both the *on-board* sub-system that is located on the train, and the *wayside* sub-system located on the tracks. Both of them receive information from external components also known as the *environment* of the system. The on-board subsystem controls doors opening and closing, braking, train positioning, train speed and stop as well as broadcasting information to the passengers. While the wayside subsystem mainly delivers movement authorizations according to the train speed and position. As a running example, we present the speed dependent door supervision scenario. It consists of the train doors opening supervision by the train sub-systems, according to the train speed. Hence, the train is allowed to open the doors if and only if the train reaches a minimum speed. It must supervise continuously the train speed.

We note *S* the speed dependent door supervision scenario. The description of *S* in the functional specification operational scenarios starts with the following informal description:

> *The train is fully berthed at a platform and stops. The train doors are released and opened. The train starts to roll away. The emergency brake is applied when the train exceeds a certain minimum speed.*

This description consists of a series of assertions. The first assertion is the prerequisite for the doors opening; the train must arrive at a platform and stop (reach a minimum speed) to release the train doors. After that, the train starts supervising the doors. If the train starts to roll away and does not exceed a minimum speed then the doors release is maintained. Otherwise, the train applies an emergency brake and revokes the door release. The scenario description continues with the definition of the initial conditions using Table 1.

The objective of this table is to define the necessary conditions for *S*. However, there are some missing information. The wayside is responsible for the train movement authority, but is not defined in the initial conditions. The initial conditions are also very detailed with the use of data telegrams (op_train_sup_limit_low), and at the same time not very clear with the use of acronyms (SM-CTC).

Table 1 Initial conditions

Component/Sub-system	Initial conditions
On-board sub-system	In SM-CTC
	With speed-dependent door supervision (open doors ignored at low speed, op_train_sup_limit_low > 0)
Platform	Without PSDs

Table 2 Speed dependent supervision scenario

Step	Action/Event	Comment
1	The train approaches the stopping point, it is already fully berthed. The on-board subsystem indicates this to the HMI (via HMI_O_In_Stopping_Window) and to the TMS (via PIS_O_Fully_Berthed_Side_Indication)	#REQ-AS_TGMT_R2-platform_stopping_window-01# #REQ-AS_TGMT_R2-fully_berthed_indication-01#
2	The train comes to a standstill. The on-board subsystem releases the train doors doors_release-01# at the correct side(s) via TCL_O_Door_Release_Left/Right	#REQ-AS_TGMT_R2-
3	The driver initiates door opening via CAB_I_Door_Open_Command. The on-board subsystem opens the train doors via TCL_O_Opening/Closing_Doors_Left/Right	#REQ-AS_TGMT_R2-manual_door_mode-01#
4	The doors open. This is reported to the on-board subsystem via TCL_I_Door_Closed_Indication	
5	The on-board subsystem indicates the open doors to the HMI (via HMI_O_Train_Door_Status). It sets the recommended speed to zero (HMI_O_Recommended_Speed) and sets the EBIC speed to op_train_door_sup_limit_low_tp_speed_err_model (HMI_O_EBIC)	#REQ-AS_TGMT_R2-train_door_indication-01# #REQ-AS_TGMT_R2 -door_supervision_HMI-01# #REQ-AS_TGMT_R2 -door_supervision_HMI-03#
6	The train starts to move (standstill window with op_max_movement_distance left), the configured minimum speed for the door supervision (op_train_door_sup_limit_low) is not yet exceeded. The on-board subsystem reacts by revoking the door release (via TCL_O_Door_Release_Left/Right)	#REQ-AS_TGMT_R2-doors_release-01#
7	While rolling, the train loses the fully berthed status. The on-board subsystem revokes the fully berthed indication to the HMI (HMI_O_In_Stopping_Window) and to the TMS (PIS_O_Fully_Berthed_Side_Indication)	#REQ-AS_TGMT_R2-platform _stopping_window-01# #REQ-AS_TGMT_R2-fully _berthed_indication-01#
8	The train exceeds the configured minimum speed for the door supervision (op_train_door_sup_limit_low). The on-board subsystem applies and emergency brake (TCL_O_Emergency_Brake)	#REQ-AS_TGMT_R2-train _door_supervision_reaction-01#

After that, the scenario is described in details (see Table 2). It refers to requirements, components, communication channels and interfaces.

Even if the scenario refers to different information, there is still some remaining implicit one. For example, the components allocated to each action are not precised (use of passive voice in step 4). The execution order of actions is also implicit. For example, it is not precised whether the train is continuously under supervision or after the train stop. Moreover, actions may include components not mentioned in the initial conditions table. For example, for the action: "*The* ***on-board subsystem*** *indicates this to the* ***HMI***". None of the components here in bold are precised in the initial conditions of Table 1. Moreover, actions may not be defined at a specific abstraction level of the system architecture. For example, the scenario may include an action of the train and an action of the on-board sub-system, which is a sub-system of the train.

The scenario displayed in Table 2 is part of the operational scenarios document regrouping such similar scenarios. It means that modeling such document requires the study and modeling of 104 of those scenarios. The study of a scenario starts with its thorough study of it and some discussions with experts to complete the information we have. Therefore, we retrieve components and actions and thrive ambiguities.

As mentioned in Sect. 1, we investigated the use of SysML with its diagrams dedicated to the scenarios representation as activity or sequence diagrams. The technical concepts of the scenarios document have no direct representation, and the use of different views lacks coherence. Moreover, our objective is to capture the maximum of information in a model. Graphics, whether using SysML for example, are not sufficient to represent the total amount of information. Therefore, a textual model always needs to be used as a reference. Thus, the idea is to define a modeling language tailored to the TGMT railway system concepts with both textual and graphical representations.

3 *ScOLa*

ScOLa is a modeling language dedicated to the formalization of system specifications. It models the structure of the system and its behavior by means of scenarios.

The system structure in *ScOLa* is represented using two viewpoints. The first one is the architecture of the system; a hierarchy of components. A component is a physical or structural constituent of the system. For example, the system *TGMT* is a component and it is composed of sub-components that are the *on-board*, the *wayside*, the *on-board* environment and the *wayside* environment. Thus, the system architecture of a model in *ScOLa* is a *hierarchy* of components. Its description requires that every sub-component has to belong to one and only one component. One way of describing this hierarchy would be the use of an object-oriented paradigm using the concept of class. However, there might be cases in the system behavior where the concept of external component intervenes. This concept means that we can define components without any relation to their upper components, the so-called *prototypes* [6]. The description of the system structure in *ScOLa* is then *prototype-oriented*. The second viewpoint of the system structure is the concept of *block*. It represents a set of components aggregated from the system architecture viewpoint. A block can depict the components involved in several scenarios and a scenario uses only one block.

A *ScOLa* model is a set $\mathcal{S}$ of scenarios that describe the system behavior at different abstraction levels. A scenario can be decomposed into sub-scenarios or a set of atomic actions $\mathcal{A}$. They are realized by a set of components $\mathcal{C}$, either individually or in cooperation.

Definition 1 A *ScOLa* model $\mathcal{M}$ is defined by the tuple $< \mathcal{S}, \mathcal{A}, \mathcal{C}, \mathcal{L}, \mathcal{O}p >$ where:

- $\mathcal{S}$ is the finite set of scenarios that describe the behavior of the system;
- $\mathcal{A}$ is the finite set of atomic actions the scenarios are built of;
- $\mathcal{C}$ is the set of physical components that build the structural architecture of the system;
- $\mathcal{L}$ is the set of possible abstraction levels of the system;
- $\mathcal{O}p$ is the finite set of operators, where $\mathcal{O}p = \{$precedence, parallelism, preemption, refinement$\}$.

(1) Concept of component

A component $c \in \mathcal{C}$ is defined by the tuple $< Id_C, \mathcal{A}(c), \mathcal{C}(c), L_c >$ where:

- Id_C is the unique identifier of c;
- $\mathcal{A}(c)$ is the set of actions allocated to component c, $\mathcal{A}(c) \subset \mathcal{A}$;
- $\mathcal{C}(c)$ is the set of the components children if it applies, empty otherwise, with $\mathcal{C}(c) \subset \mathcal{C}$;
- L_c is the level of abstraction where the component is defined.

A component c is said *complex* when it can be decomposed into sub-components, otherwise it is said *basic*.

(2) Concept of scenario

A scenario describes a step in the system behavior. It may aggregate sub-scenarios or *actions*. Each scenario describes a partial view of system behavior. A scenario $s \in \mathcal{S}$ is defined as $s =< Id_s, L_s, \mathcal{F}(s) >$ where:

- Id_s is the unique identifier of the scenario;
- L_s is the abstraction level of scenario s;
- $\mathcal{F}(s)$ is the set of scenarios or actions encapsulated in s, with $\mathcal{F}(s) \subset \mathcal{A} \cup \mathcal{S}$.

(3) Concept of action

An action $a \in \mathcal{A}$ is by definition an atomic scenario. It is defined using the following tuple $a =< Id_a, \mathcal{C}(a), L_a, \mathcal{T}(a) >$ where:

- Id_a is the unique identifier of a;
- $\mathcal{C}(a)$ is the set of components realizing a;
- L_a is the abstraction level of a, $i \in \mathbb{N}$;
- $\mathcal{T}(a)$ is the corresponding type of the action.

Indeed, as a may require input data and/or produce results, it may be of one of the following types:

- *Simple action* when it requires the resources of a single component to be completed. This type of action may require input data, that may be provided by one or several other actions. The input data, if there are any, are analyzed in order to generate an output result, after some process and calculation. Formally, let $\mathcal{A}_s$ be the set of *simple actions*. If $a \in \mathcal{A}_s$, then $\exists c \in C$ such as $a \in \mathcal{A}(c)$.
- *Transfer action* when an action is shared between two or more components. Such an action can be a data transmission between two components of the system, and thus requires the *cooperation* of both components. Let $\mathcal{A}_t$ be the set of *transfer actions*. If $a \in \mathcal{A}_t$ then $\exists c_1, c_2 \in C$ such as $a \in \mathcal{A}(c_1) \cap \mathcal{A}(c_2)$.
- *Question action* when a allows the system to choose between two or more alternative behaviors. Typically, a question action can be a test on data in order to choose which scenario to proceed within the next step. Let $\mathcal{A}_q$ be the set of *question actions*. If $a \in \mathcal{A}_q$ then $\exists a_1, a_2, \ldots, a_n \in \mathcal{A}$ such that executing a leads to the execution of a_1 *or* a_2 or ... or a_n.

(4) Concept of refinement

Because the different views of the system architecture may provide too detailed functions (functional view), components (organic view) and events (event-based view), it becomes necessary, during the system engineering process, to structure these information and introduce a certain hierarchy between them. Thus, we use the notion of *refinement* as one of our main language concept. The refinement of a scenario s of abstraction level l_n is a set of sub-scenarios $s_1, s_2, \ldots, s_k$, $k \in N$, of abstraction level l_{n+1}. It is represented using encapsulation.

(5) Concept of Precedence

It models the sequential completion of the actions or scenarios. If a_1,$a_2 \in \mathcal{A}$, a_1 and a_2 follow a precedence order, noted $a_1 \to a_2$, if a_2 needs to wait for the completion of a_1 in order to proceed.

(6) Concept of Parallelism

It models the independence in the actions or scenarios realization. If $a_1, a_2 \in \mathcal{A}$, a_1 and a_2 are processed in parallel, noted $a_1 || a_2$, if the order of execution is meaningless. In *ScOLa*, parallelism represents a particular case of precedence where a_2 may proceed before a_1 or a_1 may proceed before a_2.

(7) Concept of Preemption

It models the choice between two actions or scenarios of the system. Given a question action $a \in \mathcal{A}_q$, such that $a \to (a_1 + a_2)$ where a_1 and $a_2 \in \mathcal{A}$,

$$\begin{cases} a \text{ is followed by } a_1 & \textit{if a is true} \\ a \text{ is followed by } a_2 & \textit{otherwise} \end{cases}$$

Representation of *ScOLa* models

ScOLa proposes both textual and graphical modeling of a system. The following table (Table 3) presents the language textual and graphical representations.

Table 3 Textual and graphical representations of *ScOLa* concepts

Operator	Graphical representation	Textual representation
Scenario	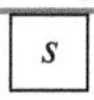	**Scenario**
Simple action		**Action** a **by** c
Transfer action	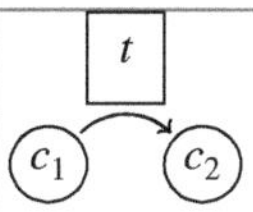	**Transfer** t **from** c_1 **to** c_2
Question action		**If** (Q) **else**
Component		**Component**, **Basic-component**
Parallelism	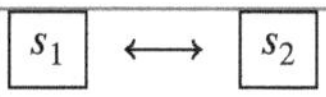	$s_1 \parallel s_2$
Precedence	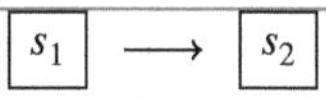	$s_1 \rightarrow s_2$

4 Incremental Modeling of the System Specification Scenarios

We present an incremental methodology to model the running example scenario *S* and its architecture, using the graphical and textual representations of *ScOLa*.

4.1 Incremental Modeling of the Running Example

We start with the definition of the architecture on which *S* takes place (see Fig. 1). As mentioned in Sect. 2, the TGMT is composed of the on-board and wayside sub-systems. It is also composed of the on-board and wayside environments. If driver-with, the system also involves a driver. In *S*, the OBCU (On-board Communication Unit), sub-component of the on-board, controls and monitors the train movement. The HMI (Human Machine Interface), sub-component of the on-board environment, is used as a communication mean between the OBCU and the driver. The ATS (Automatic Train Supervision), sub-component of the wayside environment, monitors the trains and adjusts the performance of trains individually. From the initial conditions table of *S* (see Table 1), we get that it is realized by the on-board sub-system and the platform. However,as we have seen, the wayside sub-system is also involved. In particular,its environment sub-component. The HMI, is also involved as well as a driver. Moreover, there is no mention of the platform in the scenario description. Therefore, the instances of the architecture involved in *S* are the ones depicted in Fig. 1. Once the components of the scenario clearly defined, we can start the incremental model-

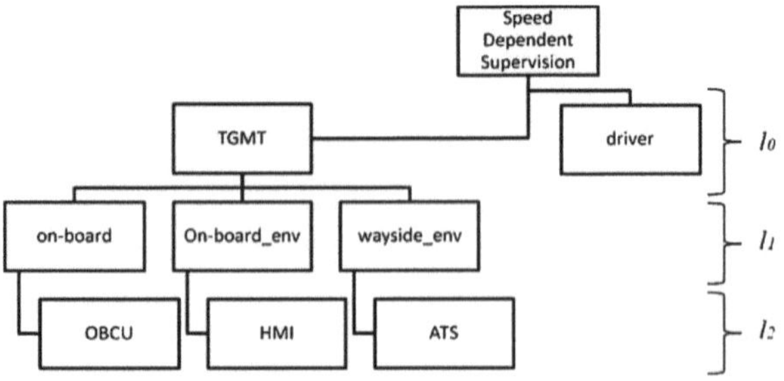

Fig. 1 Architecture of the TGMT system involves in S

ing of S at several abstraction levels. Since the system architecture has three layers, we model S at three abstraction levels noted l_0, l_1 and l_2.

S at abstraction level l_0 consists in a series of sub-scenarios realized by the TGMT. Such an abstraction allows having a wider and easier understanding of what the scenario is doing, and is useful at the system level test and safety analysis. A scenario at abstraction level l_0 does not depict the internal behavior of sub-components of the train. In the system specification description (see Table 2), steps of S depict components at several abstraction levels. In order to define the abstraction level l_0, we provide the following assertions on each step of S:

- We skip all internal communications between the TGMT sub-components.
- The main event of Step 1 is to be fully berthed when arriving at the stopping point. It is followed by an internal communication between sub-components of the TGMT.
- Step 2 consists in the doors release at the correct side.
- The TGMT opens the train doors in Step 3.
- The TGMT detects that the doors are indeed open in Step 4.
- The TGMT sets the speed to zero in Step 5.
- Step 6 provides two behaviors: the TGMT detects whether the train minimum speed is exceeded and revokes the door release.
- Steps 7 and 8 are about testing whether minimum configured speed is reached. The TGMT applies an emergency brake if needed.

We note s_{ij} a sub-scenario of S, where i is the corresponding abstraction level and j the sequencing number in S at abstraction level l_i, $i = 0, 1, 2$. The words in bold are the sub-systems of the architecture. Therefore, the representation of S at abstraction level l_0 is depicted as follows:

s_{01}: The **TGMT** is fully berthed.
s_{02}: The **TGMT** releases the train doors at the correct side.
s_{03}: The **driver** initiates door opening.
s_{04}: The **TGMT** opens the train doors.
s_{05}: The **TGMT** detects that doors are open.
s_{06}: The **TGMT** sets the recommended speed to zero.

s_{07}: The **TGMT** detects that the configured minimum speed for the door supervision is not yet exceeded.
s_{08}: The **TGMT** reacts by revoking the door release.
s_{09}: The **TGMT** tests if the train exceeds the configured minimum speed for the door supervision.
s_{010a}: The **TGMT** applies an emergency brake.

Therefore, S at abstraction level l_0, is a model $\mathcal{M} =< \mathcal{S}, \mathcal{A}, \mathcal{C}, \mathcal{L}, \mathcal{O}p >$ where:

$$\begin{aligned}
\mathcal{S} &= \{s_{01}, s_{02}, s_{04}, s_{05}, s_{06} s_{07}, s_{08} s_{09}, s_{010a}\} ,\\
\mathcal{A} &= \{s_{03}\} ,\\
\mathcal{C} &= \{TGMT, driver\} ,\\
\mathcal{L} &= l_0 ,\\
\mathcal{O}p &= \{precedence, parallelism, choice\}
\end{aligned}$$

The abstraction level l_1 of S uses more detailed level of the system architecture: the on-board, the wayside, the on-board_env and the wayside_env. At this stage, we propose more detailed descriptions of the steps of the scenario. The added value lies in the interactions between the train sub-components and the allocation of some actions to these sub-components. For example, consider Step 1 of S where the train is detected to be fully berthed. The sub-component responsible for this action is the on-board. Therefore, l_1 description of the step 1 would rather be s_{11}: The **on-board** detects that the train is fully berthed. This information comes from the different documents of the system specification. The original definition was ambiguous with the use of passive voice and the non-allocation of actions to the correct sub-components. Hence, the refinement of S from abstraction level l_0 to l_1 requires the following additional information:

- The on-board controls the train doors release and opening via instructions from the wayside_env.
- The wayside_env is the sub-system responsible for the train supervision. Thus, it is responsible for detecting the train speed and communicating the information to the on-board and finally, the on-board_env initiates the doors opening.

It results the following representation of S at abstraction level l_1:
s_{01}: —s_{11}: The **on-board** detects that the train is fully berthed.
—s_{12}: The **on-board** reports the fully berthed information to the **on-board_env**.
s_{02}: —s_{11}: the **on-board** releases the train doors at the correct side.
s_{03}: —s_{11}: The **driver** initiates door opening.
s_{04}: —s_{11}: The **on-board** opens the train doors.
s_{05}: —s_{11}: The **wayside_env** detects that doors are open.
—s_{12}: The **wayside_env** reports the information to the **on-board**.
s_{06}: —s_{11}: The **on-board** indicates the open doors to the **on-board_env**.
—s_{12}: The **on-board** sets the recommended speed to zero.
s_{07}: —s_{11}: The **wayside_env** detects that the configured minimum speed for the door supervision is not yet exceeded.

—s_{12}: The **wayside_env** transfers the information to the **on-board**.
s_{08}: —s_{11}: The **on-board** reacts by revoking the door release.
s_{09}: —s_{11}: If the **wayside_env** detects that the train exceeds the configured minimum speed for the door supervision.
s_{010a}: —s_{11}: The **on-board** applies an emergency brake.
$\mathcal{M}$, the model associated with S at abstraction level l_0 is refined into sub-models at abstraction level l_1. Each sub-model is associated with a sub-scenario of S.

As for the previous refinements, definition of S, at abstraction level l_2 requires the allocation of the actions of abstraction level l_1 to l_2 sub-components. Provided that the ATS, HMI and OBCU are the only sub-components of the wayside_env, on-board_env and on-board respectively involved in S, the scenario is depicted at abstraction level l_2 as follows:
s_{01}, s_{11}: —s_{21}: The **OBCU** detects that the train is fully berthed.
s_{01}, s_{12}: —s_{21}: The **OBCU** indicates this to the **HMI**.
s_{02}, s_{11}: —s_{21}: the **OBCU** releases the train doors at the correct side.
s_{03}, s_{11}: —s_{21}: The **driver** initiates door opening.
s_{04}, s_{11}: —s_{21}: The **OBCU** opens the train doors.
s_{05}, s_{11}: —s_{21}: The **ATS** detects that doors are open.
s_{05}, s_{12}: —s_{21}: The **ATS** reports the information to the **OBCU**.
s_{06}, s_{11}: —s_{21}: The **OBCU** indicates the open doors to the **HMI**.
s_{06}, s_{12}: —s_{21}: The **OBCU** sets the recommended speed to zero.
s_{07}, s_{11}: —s_{21}: The **ATS** detects that the configured minimum speed for the door supervision is not yet exceeded.
s_{07}, s_{12}: —s_{21}: The **ATS** transfers the information to the **OBCU**.
s_{08}, s_{11}: —s_{21}: The **OBCU** reacts by revoking the door release.
—s_{22}: The **OBCU** revokes the fully berthed indication to the **HMI**.
s_{09}, s_{11}: —s_{21}: If the **ATS** detects that the train exceeds the configured minimum speed for the door supervision.
s_{010a}, s_{11}: —s_{21}: The **OBCU** applies an emergency brake.
In addition to this formalization of the specification scenario, we propose the graphical representation of S as depicted in Figs. 2 and 3. Due to the space limitation on this paper, the textual representation of S is depicted in Annex. A.

4.2 *Incremental Modeling of the System Specification Document*

The operational scenarios specification includes 104 scenarios. The initial conditions tables involve an average of 5 components from the system architecture with a minimum of 2 and a maximum of 10 components. Note that this metric is rather meaningless since we notice that often components are explicitly involved in the scenario description without any prior definition in the initial conditions table.

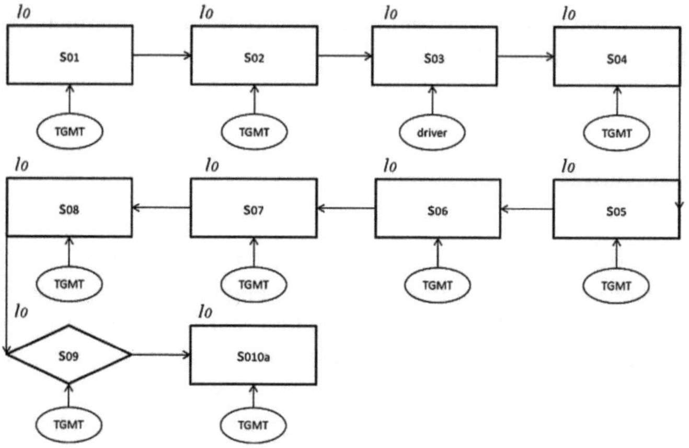

Fig. 2 Graphical representation of S at abstraction level l_0

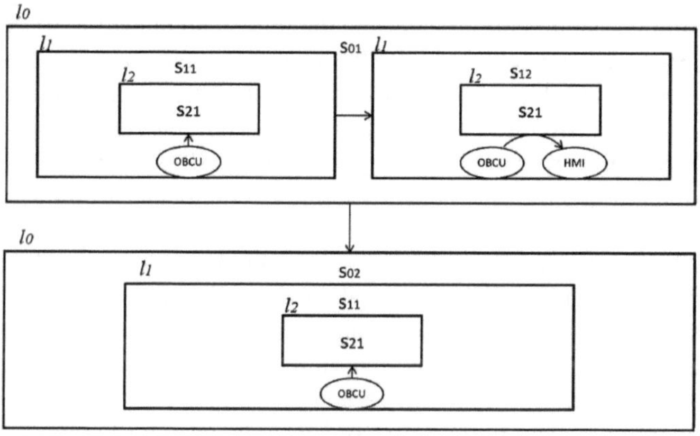

Fig. 3 Partial graphical representation of S at all abstraction levels

Scenarios tables include an average of 9 steps with a minimum of 3 and a maximum of 16. Once again, as depicted in Table 2, these steps do not describe a single behavior but rather an average of 3 to 4 behaviors. A single behavior depicts an action realized by a single component or a communication between two or several components. Among the single behaviors retrieved from the steps, only 2/3 of them have a clear structure: 2/3 relate an action realized by a single component and less than a third represent a communication between two or several explicitly defined components. The remaining actions represent test actions. Besides, the remaining third of the single behaviors have an ambiguous description: half of them express a communication action but the components are implicit while the other half are meaningless sentences.

Therefore, the formalization of the operational scenarios document requires the analysis of around 2800 actions. These scenarios represent at least, an important part of the behavior of the system. It is an interesting communication means for engineers and helps the ambiguity resolution.

5 Related Work

A MBSE methodology is characterized as a set of related processes, methods and tools used to support the discipline of systems engineering in a "model-based" or "model-driven" context. MBSE has been addressed in literature by two types of approaches, semi-formal and formal. Semi-formal methods rely on graphical notations to represent different aspects of the system, by means of specialized views. The graphics are used as a communication support between stakeholders. It allows a wider view of the system context than existing textual specifications. Throughout the years, several MBSE methodologies and tools were proposed. OOSEM [7] of INCOSE [8] integrates a top-down model based approach, analyzing system stakeholders, defining system requirements and the logical architecture. Rational Unified Process for System Engineering (RUP SE) [9] of IBM is an iterative methodology for the system design with four phases: inception, elaboration, construction and transition. The ARCADIA (ARchitecture Analysis and Design Integrated Approach) [10] is a model-based engineering method for systems design. It is developed by Thales and relies on a domain specific language providing several means for the system, logical and physical architecture representation of a complex system.

SysML [2], a semi-formal graphical notation standardized by the Object Management Group (OMG) [11], provides several views or diagrams to represent the architecture of a system (Block Definition Diagrams for hierarchical views and Internal Block Diagrams for architecture internal views). It also provides several views for the behavior representation of systems, among which use case diagrams, activity and sequence diagrams, ... etc. In order for SysML to be used for the modeling of complex systems, a methodology is required. We distinguish two types of use of SysML artifacts. The first one is the use of profiles [2], that is to specialize the language for the experts needs. It is a flexible way to only represent the needed views of the system. SysML itself is a UML [12] profile for modeling complex systems instead of complex software. SysML profiling is used in the industry. Valeo with SysCars [13] propose a subset of SysML diagrams and a methodology to sequence the modeling activities to be performed. However, such method requires a deep knowledge of the language. Moreover, there might be no isomorphism between the concepts in the system specification and the targeted modeling language, which must be adapted. The second type of SysML use is the integration of viewpoints. It is a non-intrusive extension of the modeling language. ASAP [14] methodology of Alstom is a good example. This is a top-down approach with several views (operational, functional and constructional) to represent the requirements and the model of a railway system.

The requirements allocation to the model at each view allows a more systematic model validation. However, both approaches are only valid for a specific context. It considers modeling objects instead of concepts. It might result redundant information when several views contain similar information. Moreover, it does not consider the system evolution.

Formal methods are mathematically-based languages, techniques and tools. They are used for specification and verification of systems. For example, the B-method based on the B language [1] specifies and designs system softwares. Scade [15] is also a certified formal language used for system development, used in multiple domains. Other languages as AADL [16] and Uppaal [17] provide the same properties. However, the entry cost of such methods and languages is high. Formal models are intuitive but necessitate an expertise.

Model transformation builds a bridge between both worlds. There are many semi-formal to formal models transformations in literature as SysML to Uppaal [18], UML-B [19] or SysML to Altarica [20]. But, a complete coverage of both languages is difficult to obtain. Moreover, consistency issues are due to the lack of the expressiveness of formal models while semi-formal models depicts more information.

Most of existing applications of the MBSE are methodologies to model a complex system. However, there is a gap between system specification descriptions and models. Therefore, *ScOLa* proposes a reverse engineering approach to define the necessary concepts for an efficient system modeling by means of a small set of concepts.

6 Conclusion

This paper presents an incremental methodology to model system specifications by means of scenarios. The methodology is based on the scenario-based language *ScOLa*. It defines formally the concepts required for a precise system modeling. While most modeling methodologies define objects, *ScOLa* is based on a minimum set of concepts to model the system structure and behavior of a complex system. The language is based on specifications of railway systems, but we believe it can be useful for a broader range of applications. The next step is to study the use of *ScOLa* for safety analysis purposes.

A Textual Representation of *S*

The textual representation of *S* using the textual operators of *ScOLa* is depicted in Fig. 4

```
System TGMT {
    Architecture architecture {
        Component TGMT {
            Component onboard {
                Basic-Component OBCU
            }
            Component onboard_env {
                Basic-Component HMI
            }
            Component wayside_env{
                Basic-Component ATS
            }
        }
            Basic-Component Driver
        }

      Block b1 {
      architecture.TGMT TGMT;
      architecture.TGMT.onboard_env.HMI HMI;
      architecture.TGMT.onboard.OBCU OBCU;
      architecture.TGMT.wayside_env.ATS ATS;
      architecture.Driver driver;
      }

      Scenario S with b1 {
        Scenario s01 = "The TGMT is fully berthed"
        {
            Scenario s11 = "The on-board detects that the train is fully berthed"
            {
                Action s21 = "The obcu detects that the train is fully berthed" by b1.OBCU ;
                Script s21;
            }
            Scenario s12 = "The on-board indicated the fully berthed status to the on-board_env" {
                Transfer s21 = "The obcu sub-system indicates this to the hmi" from b1.OBCU to b1.HMI;
                Script s21 ;
            }
            Script s11 -> s12;
        }

        Scenario s02 = "The TGMT releases the train doors at the correct side"
        Scenario s03 = "The driver initiates door opening"
        Scenario s04 = "The TGMT opens the train doors"
        Scenario s05 = "The TGMT detects that doors are open"
        Scenario s06 = "The TGMT sets the recommended speed to zero"
        Scenario s07 = "The TGMT detects that the configured minimum speed for the door supervision is not yet exceeded"
        Scenario s08 = "The TGMT reacts by revoking the door release"
        Test s09 = "If the TGMT exceeds the configured minimum speed for the door supervision" {
        Scenario s010 = "The TGMT applies an emergency brake" }
        Script s01 -> s02 -> s03 -> s04 -> s05 -> s06 -> s07 -> s08 -> s09 -> s09.s010;
      }
}
```

Fig. 4 Textual representation of *S*

References

1. Abrial, J.-R.: The B-Book: Assigning Programs to Meanings. Cambridge University Press (2005)
2. Friedenthal, S., Moore, A., Steiner, R.: A Practical Guide to SysML: the Systems Modeling Language. Elsevier (2011)
3. Issad, M., Kloul, L., Rauzy, A., Berkani, K.: ScOLa, a scenario oriented modeling language for railway systems. INSIGHT **18**(4), 34–37 (2015)
4. White, S.A.: Introduction to bpmn. IBM Cooperation **2**, (2004)
5. Harel, D.: Statecharts: a visual formalism for complex systems. Sci. Comput. Program. **8**(3), 231–274 (1987)
6. Naumann, J.D., Jenkins, A.M.: Prototyping: the new paradigm for systems development. Mis Q. 29–44 (1982)
7. Lykins, H., Friedenthal, S., Meilich, A.: Adapting uml for an object oriented systems engineering method (oosem). In: Proceedings of the 10th Annual INCOSE Symposium, International Council on Systems Engineering (July 2000). http://www.omg.org/docs/syseng/02-06-11.pdf (2000)

8. Wiley, et al.: INCOSE Systems Engineering Handbook: A Guide for System Life Cycle Processes and Activities. John Wiley & Sons (2015)
9. Cantor, Murray: Rup se: The rational unified process for systems engineering. Ration. Edge Ration, Softw (2001)
10. Voirin, J.-L., Bonnet, S.: Arcadia: model-based collaboration for system, software and hardware engineering. In: Complex Systems Design & Management, poster workshop (CSD&M 2013) (2013)
11. OMG CORBA and IIOP Specification: Object management group. Joint revised submission OMG document orbos/99-02 (1999)
12. Muller, P.-A., Gaertner, N.: Modélisation objet avec UML, vol. 514. Eyrolles Paris (2000)
13. Piques, J.D., Andrianarison, E.: Sysml for embedded automotive systems: lessons learned. Interfaces **3**, 3b (2011)
14. Góngora, H.G., Ferrogalini, M., Moreau, C.: How to boost product line engineering with mbse-a case study of a rolling stock product line. In: Complex Systems Design & Management, pp. 239–256. Springer (2015)
15. Abdulla, P.A., Deneux, J., Stålmarck, G., Ågren, H., Åkerlund, O.: Designing safe, reliable systems using scade. In: Leveraging Applications of Formal Methods, pp. 115–129. Springer (2006)
16. Feiler, P.H., Gluch, D.P., Hudak, J.J.: The architecture analysis & design language (aadl): an introduction. Technical report, DTIC Document (2006)
17. Larsen, K.G., Pettersson, P., Yi, W.: Uppaal in a nutshell. Int. J. Soft. Tools Technol. Transfer (STTT) **1**(1), 134–152 (1997)
18. De Saqui-Sannes, P., Apvrille, L., et al.: Avatar/ttool: un environnement en mode libre pour sysml temps réel. Génie Logiciel **58**(98), 22–26 (2011)
19. Snook, C., Butler, M.: Uml-b: Formal modeling and design aided by uml. ACM Trans. Soft. Eng. Methodol. (TOSEM) **15**(1), 92–122 (2006)
20. Cressent, R., David, P., Idasiak, V., Kratz, F.: Apports de sysml à la modélisation des systèmes complexes à fortes contraintes de sûreté de fonctionnement. In: ITT'09 (Technological Innovation and Transport Systems 2009), p. p39 (2009)

Automated Piping with Standardized Bends in Complex Systems Design

Samuel Vogel and Stephan Rudolph

Abstract Combining subsystems to build a fully integrated product is a challenging task in complex systems design. The integration of flow components requires a fast creation and validation of different pipe route variants. In this article an algorithm for the automated generation of pipe routes in a given installation space is presented. The pipe route generation is constrained to the usage of prechosen (standardized) pipe bend sets. The routes are rule-based manipulated and evolved using a simulated annealing optimization scheme.

1 Introduction

Pipe routing is a task that frequently occurs in several engineering processes. Starting from plant engineering and construction up to aircraft and spacecraft structures, the synthesis of optimal pipe routes, especially ones whose degrees of freedom are restricted to the usage of standardized pipe bends, is a challenging task. First of all, generic routing algorithms on graphs shall be considered [1]. The archetypal task of finding optimal paths on weighted graphs is known as the (shortest) path problem [2–5]. There are many routing applications that are executed on discretized geometries which can be written as graphs. Such path algorithms are used in chip design to find optimal circuit paths [6–8]. Other applications for routing algorithms can be found in robot motion planning [9, 10] or in ship route planning [11].

Especially in pipe route applications grid-based path search algorithms are used: Applications range from piping in aerospace design [12] and pipe route design in ship construction [13, 14] to architectural pipe routing in designing building services [15]. Other pipe route algorithms are incorporating knowledge-based methods, such as expert systems, to route ship pipes [16] or use multi-objective genetic algorithms for a determination of pipe arrangements [17]. These references mentioned

S. Vogel (✉) · S. Rudolph
Institute of Statics and Dynamics of Aerospace Structures,
University of Stuttgart, Stuttgart, Germany
e-mail: samuel.vogel@isd.uni-stuttgart.de

G. Fanmuy et al. (eds.), *Complex Systems Design & Management*,
DOI 10.1007/978-3-319-49103-5_9

are restricted to create paraxial, orthogonal routes. The piping algorithm shown in [18] is able to integrate non-orthogonal pipe bends to bypass obstacles. But it is still restricted to paraxial routes adjacent to the bypass.

For general purpose engineering applications, especially in tight and warped installation spaces, a synthesis of non-orthogonal and non-paraxial pipe routes is mandatory. The flow interfaces of the components that shall be linked by the pipe routes are not necessarily paraxially aligned. Such pipe routes between non-paraxial components are preferentially made up of standardized pipe bends with fixed bend angles α_i and fixed bend radii R_i to secure low manufacturing costs and to fulfill standardization requirements [19] like the *DIN 2605* [20, 21].

The above mentioned approaches are not able to combine the usage of standardized bends together with arbitrary start and end positions and directions as combining these features makes it harder to find compatible routes as the available DoF are heavily restricted. The algorithm proposed in this article follows and extends [22] and overcomes the restrictions mentioned before. It is able to synthesize pipe routes between unrestricted start and end positions and directions. This fits well into a complex systems design process where in a first step the topology of the flow network as well as the components positions are fixed and after that the components are connected by pipe routes.

2 Methods

Shape grammars are rule-based systems to generate engineering product shapes with formal synthesis techniques [23, 24]. *Antonsson* and *Cagan* are presenting in [25] a wide range of formal engineering design synthesizing systems for architectural and engineering applications that are able to produce a wide range of shapes and geometries. *Rudolph* pushes the idea of shape grammars to an even more abstract level. He proposes rule-based manipulations of a graph-based universal data model that are embedded in an adaptive production system, as a procedure to automatize the whole engineering process of designing a virtual product [26] (see Fig. 5).

2.1 Shape Grammar

The production system of a shape grammar begins with the first rule, called the *axiom*, that creates an initial shape representation [27]. The subsequent rules in the production system iteratively manipulate the shapes. *Szykman* presents in [28] a shape grammar to synthesize nonorthogonal pipe routes with arbitrary pipe bend angles. The shape grammar is coupled to an optimization algorithm (Simulated Annealing [29]) to evolve the pipe shapes. Extending this approach by pushing the bend angles to discrete values using a penalty method within the optimization process hasn't been successful as the optimization scheme was not able to find acceptable solutions within an acceptable runtime for non axis aligned start/end positions and directions.

Nevertheless, the optimization process used in this work is similar to the process of [28]: An initial pipe route that fulfills the given constraints is generated and iteratively manipulated within a simulated annealing scheme by the *rules* presented in the sections below. A combined fitness function W is minimized. It is build by the product of the installation room intersection penalty $p_{intrsct}$ and the distance penalty $p_{dst2ref}$ of the pipe route to a given reference path:

$$W = p_{intrsct} \cdot p_{dst2ref} = \min!. \tag{1}$$

2.2 Simulated Annealing

Simulated annealing employs a fitness function W that is minimized (maximized) during the optimization cycles [29]. The Simulated Annealing algorithm works iteratively and creates a new configuration S_{i+1} based on the current system configuration S_i. The newly generated state S_{i+1} is accepted if its fitness value $W(S_{i+1})$ is lower (higher) than the fitness value of the current state $W(S_i)$. Additionally, the state S_{i+1} can be accepted with the probability

$$p(\Delta W) = e^{-\frac{\Delta W}{T}}, \tag{2}$$

with $\Delta W = W(S_{i+1}) - W(S_i) > 0$, if the new state has a worse fitness value (for maximization correspondingly $\Delta W = W(S_i) - W(S_{i+1}) > 0$) than the current state.

During optimization the temperature parameter T is lowered, analogous to the annealing of a metallic melt that moves towards its energetic minimum state when cooled down. The temperature is lowered following a specific annealing scheme that specifies how many optimization steps are taken for one temperature level. A (relative) amount that the temperature gets lowered during the course of the optimization is specified in this scheme.

2.3 Pipework Representation

The pipework shall be exclusively made up of standard pipe bend elements. Therefore the bend angles α_i are chosen from a set A of provided pipe bend angles $\alpha_i \in A = \{\alpha_0, .., \alpha_i, ..\alpha_N\}$. Each bend with angle α_i has one assigned bend radius R_j: $\alpha_i \rightarrow \{.., R_j..\}$. The straight pipe elements that are connecting the standard pipe bends have arbitrary lengths.

For reasons of simplicity, a polygonal line as an analogous model for the pipe routes is introduced (Fig. 1). The polygonal line model consists of line segments. Each line segment represents a straight pipe. The joint of two adjacent line segments

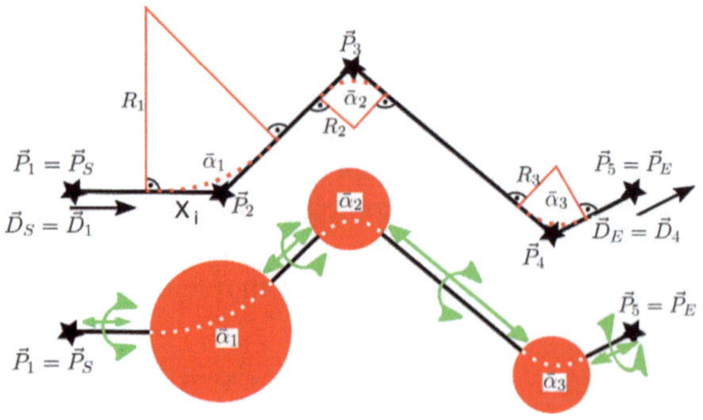

Fig. 1 Polygonal analogous model (*top*) and degrees of freedom (*bottom*) of a pipework

represents the apex of the middle line of the pipe bend. The angle between the direction vectors of adjacent line segments are representing the bend angles.

The pipe bend representation in the polygonal analogous model has to consider the space requirements of real pipe bends. Based on Fig. 1 (top) Eq. 3 can be derived to calculate the space requirement X_i of bend i.

$$X_i = tan\left(\frac{\alpha_i}{2}\right) \cdot R_i \ . \tag{3}$$

2.4 Parametric Variations

First of all, a valid pipe route made up of standard pipe bends with given start/end positions and directions is assumed. The available degrees of freedom are:

1. Straight pipe segments with variable lengths between the bends (Fig. 1 bottom, straight arrows).
2. Rotation of the pipe bends around the adjacent line segments' axes (Fig. 1 bottom, circular arrows).

Figure 1 shows these degrees of freedom as green arrows for the shown polygonal line model of the pipe route with given start and end boundary conditions. The interactions of the degrees of freedom of the pipe route with the fixed ends boundary conditions can be considered as a kinematic chain with fixed ends.

To model the allowed movement of the pipe route, within the above given movement constraints, a rigid body simulation framework is used [30, 31]. The pipe route can be modeled in the rigid body simulator as a chain arrangement of piston joints. Piston joints connect two bodies and have two degrees of freedom: A translation of the two bodies along the connecting axis and a relative rotation of the two bodies around the translation axis. This coincides with the degrees of freedom of the pipe route shown in Fig. 1. Each pair of adjacent pipe bends is represented by two bodies in the rigid body simulator that are connected by one piston joint.

To consider the space requirements of the pipe bends calculated in Eq. (3), the pipe bends i can be represented as rigid spheres with radius X_i within the rigid body simulator. The rigid body simulator [31] used contains a collision detection. Therefore the collision detection implicitly guarantees the compliance with the space requirement conditions of the pipe bends.

Parametric manipulations (via the rule *move bend*) can then be conducted within the rigid body simulator. Directed forces can be put on the rigid spheres to manipulate the pipe route within the set constraints originating from the joints, the fixed start and end boundary conditions as well as the space requirements of the pipe bends. The manipulation rule *move bend* leaves the number of pipe bends, the pipe bend angles and the bend radii unchanged. An additional polyline can be used as reference path for the parametric manipulation (*move bend*): The directed forces are applied on the pipe bends in order to push the pipe towards or onto to the reference path.

2.5 Configurational Variations: KKF

In this section a geometric construction (KKF[1]) is presented that enables the modification of the pipe bend configuration of a pipe route. The algorithm will be used to conduct all the manipulation rules that incorporate a change in the pipe bend configuration (all rules but *move bend*).

The KKF construction can be used to calculate the position of a pipe bend $i_|$ with bend angle $\alpha_{i_|}$ that is added into the pipe route between the pipe bends $i_| - 1$ and $i_| + 1$ with bend angles $\alpha_{i_|-1}$ and $\alpha_{i_|+1}$. The compliant positions of the apex of the newly introduced pipe bend can be constructed considering the surfaces of two cones with its vertices in the apex points of the two adjacent pipe bends $i_| - 1$ and $i_| + 1$. The cone axis is given by the adjacent line segment of the existing pipe route. The cone half angles are $\alpha_{i_|-1}$ and $\alpha_{i_|+1}$. This ensures the compliance with the bend angles of the pipe bends adjacent to the introduced pipe bend as shown in Fig. 2 right.

The third constraint to be fulfilled by the introduced pipe bend is to guarantee the bend angle of the newly inserted pipe bend $\alpha_{i_|}$. The positions of the pipe bend's apex that achieve this condition can be constructed using the *inscribed angle theorem* [32] known from the geometry of the plane. Using this theorem an arc between A and C represents all positions of B so that the angle $\angle ABC = \beta = 180° - \alpha_{i_|}$ leads to the required pipe bend angle $\alpha_{i_|}$ (upper left part Fig. 2).

This arc can be rotated around the line $\overline{AC}$: The resulting surface represents all positions B in 3D space connecting A and C with straight lines forming a given angle (lower left part Fig. 2). The valid positions of a new pipe bend within a given pipe route can be constructed putting the three constraints together by calculating the intersection of the two cones and the rotated arc (Fig. 2 right).

[1]This abbreviation is based on the German title *Kegel*=cone/*Kegel*=cone/*Faßkreisbogen*=inscribed angle.

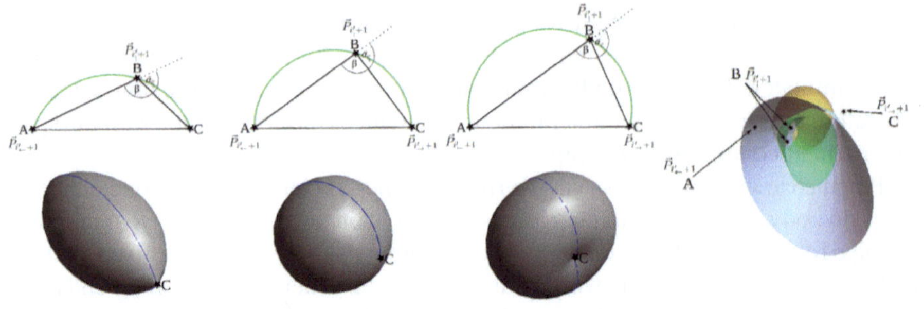

Fig. 2 *Left*: Rotated arcs (*Faßkreisbogen*) to construct an intermediate point with adjacent line segments that join in a given angle (from *left* to *right*: $\alpha_i < 90°$, $\alpha_i = 90°$ and $\alpha_i > 90°$). *Right*: KKF construction: Calculating compliant intermediate positions of the newly introduced bend for given pipe bend angles by an intersection of two cones (*green*, *grey*) and the rotated arc (*brown*) of the inscribed angle theorem

2.6 Configurational Variations: Rules

The KKF construction is used to realize the manipulation rules that conduct a modification of the pipe bend configuration. The KKF construction is practically implemented using a CAD engine to perform the intersection of the cones and the rotated arc within seconds. The rule *add bend* is realized using the KKF construction. In addition to the construction rules given above, a check is performed to test whether the newly found positions are in compliance with the space requirements X_i of the involved pipe bends. If not, the rule is rejected. If more than one possible position for a new pipe bend is found, one position is randomly chosen within the optimization procedure.

To replace an existing pipe bend with another pipe bend (via the rule *change bend*) the KKF construction can be used as well: The bend that shall be replaced is deleted and the KKF construction is used in the sense of the *add bend* rule given above. If no position for the replacement bend could be found, the original configuration is used as back-up and the rule is rejected.

The rule *remove bend* is using the KKF construction too. A first approach is to remove the bend from the line and use the KKF construction for the special case of the middle bend angle carrying the value $\alpha_{i_l} = 0°$. At the same time, the original bend angles of the bends adjacent to the originally removed bend can be recovered using the KKF construction. A second approach is to remove the bend and restore the three angles of the two preceding and the following bends (or vice versa) by using the KKF construction. To modify the first or the last bend of the pipe route, the rules presented above can be used in a slightly different manner: The bend that shall be modified is used as the first bend (or the last bend) within the bend triple of the KKF construction.

The rules guarantee to maintain a valid pipe route configuration when performing the configurational variations. A valid pipe configuration is assumed at the beginning. This initial configuration is determined using the rigid body simulator (*axiom/initialize*): A pipe route with an arbitrary pipe bend configuration that starts at the given start position with the given start direction is generated in the rigid body simulator. The free end of this pipe route is pushed towards the end position, using a force directed movement as in the *move bend* rule. When reaching the end position, the translational degrees of freedom of the free end are locked. Finally, the last pipe bend is pushed sidewise, using again a force directed approach, until the end direction condition is fulfilled. Thereafter the rotational degrees of freedom of the former free end are locked. If this conditions can not be fulfilled, e.g. due to geometrical reasons, the procedure can be repeated with a different initial pipe bend configuration. Otherwise, the simulated annealing process can be started to optimize the pipe route by iteratively applying a randomly chosen manipulation rule.

3 Results

The proposed piping algorithm can be used in many ways: Either as a stand-alone application in a wide range of engineering applications like aerospace, automotive or ship building as shown in Fig. 4 or as part of an automated design process.

3.1 Enriched Piping: Integration in Design Languages

Figure 4 shows results of the piping. The reference path in the middle of Fig. 4 has been generated using the Dijkstra routing algorithm [3] on a discretized volume model of the installation space. This reference path is used in the *move bend* parametric manipulation rule as explained above. Figure 3 shows the course of the fitness values during the synthesis of the pipe route shown in Fig. 4 left. The typical evolution of the Simulated Annealing procedure is clearly shown: Strong fluctuations of the fitness value in the beginning of the optimization for high temperature values. Decreasing fluctuations towards the end of the optimization process when reaching optimal fitness values for lower temperature levels. The runtimes of the pipe work generation shown in Fig. 3 are about to take hours. This long duration is usually not acceptable for an optimization run. The long duration is mainly caused by the absolute number of pipe elements in the pipe route as well as by shape restrictions of the available installation spaces as shown in the following section.

Second, the parallel piping shown on the right side of Fig. 4 has been created within a design language that allows the fully autonomous generation and optimization of exhaust aftertreatment systems [22]. A collision detection has been used within the rigid body simulator to handle mutual intersections between parallel pipes.

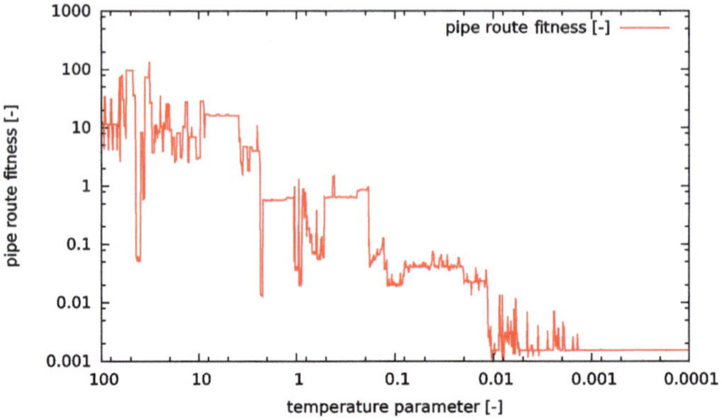

Fig. 3 Pipe route fitness versus Simulated Annealing temperature parameter of the example in Fig. 4 during the pipe route optimization process

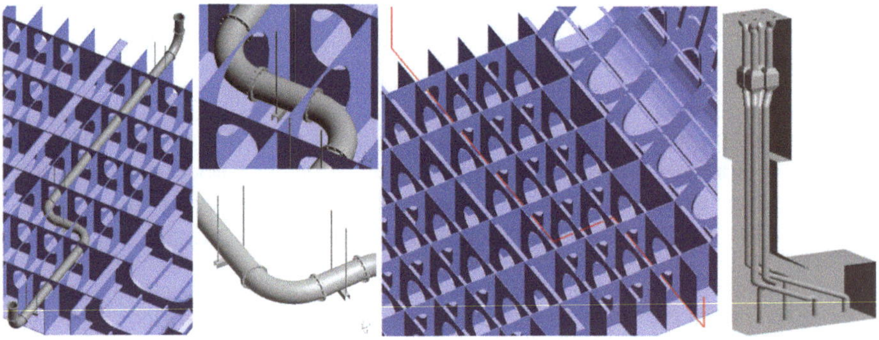

Fig. 4 Example applications: Piping and reference path within a ship body with automatically generated flanges and mounts (four pictures on the *left*). Piping in a design language for exhaust systems (*right*) [22]

Figure 5 shows the information architecture of a design language [26, 33]: Rules and vocabulary are used to create a virtual and executable image of the design process. Based on given product requirements the design language is executed by a design compiler that iteratively expands a virtual model of the product. CAE models are automatically created and executed to gather (physical) information within the virtual design and optimization process. The presented algorithm is called within the CAD plugin on the upper right in Fig. 5. The positions and directions of the generated pipe elements are sent back to the production system and allows an rule-based generation of the flanges and mounts of the piping as shown in Fig. 4 (left).

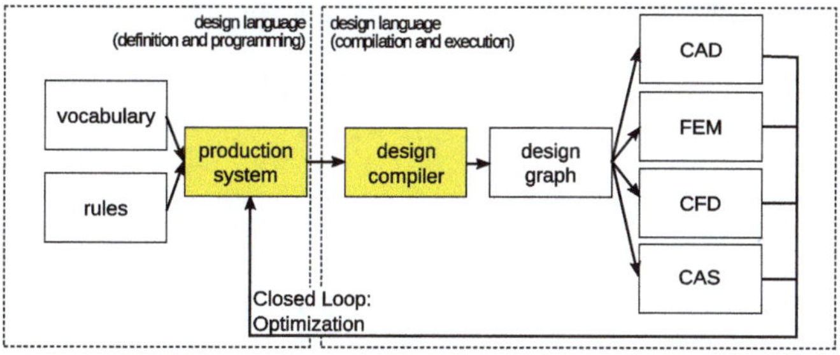

Fig. 5 Information architecture of design languages and their automatic process chains [26] (Computer-Aided Design (CAD), Finite Element Method (FEM), Computational Fluid Dynamics (CFD), Computer Algebra System (CAS))

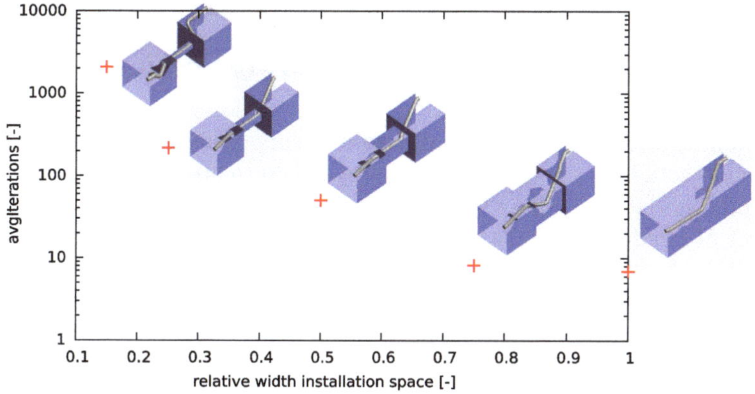

Fig. 6 Piping through installation space: Rising computing cost for finding the first valid pipe in tightening installation space

3.2 *Runtimes Versus Installation Space Complexity*

To examine the dependence of the piping algorithm's execution time on the installation space complexity, pipe routes through a simplified installation space have been synthesized. The installation spaces contain a bottleneck with different cross sections. The required number of iterations for finding a valid pipe route through the installation spaces are plotted versus the bottleneck widths in Fig. 6. The start and end positions and directions are equal for all installation spaces. It is clearly shown that the number of iterations increases rapidly with an increasing installation space complexity caused by the tightening of the bottleneck. Typical installation spaces, as shown in Fig. 4, especially the ship body in the left, contain a lot of narrow areas in the breakthroughs of the bulkheads. This complexity in combination with a significant pipe length requires an alternative approach of synthesizing the pipe routes which is presented in the next section.

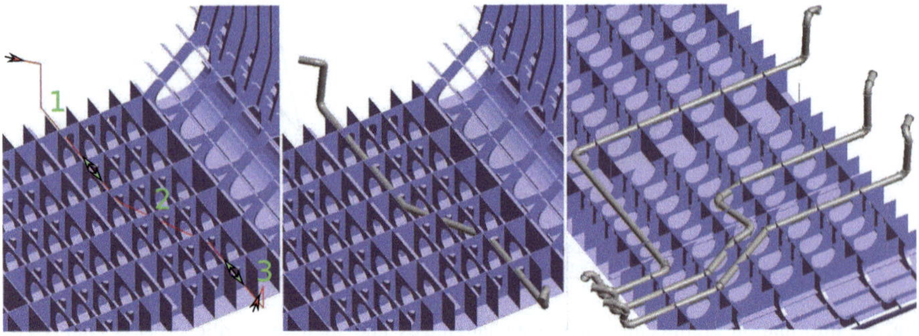

Fig. 7 Pipes generated in the ship body using the divide and conquer approach with DoF-based splitting. From *left* to *right*: split reference paths, pipe route, multiple pipe routes with flanges

3.3 Acceleration by Divide and Conquer Approach

The runtime can be reduced by the application of a divide and conquer approach. The pipe route is split in to sub routes as shown by the green arrows and figures in Fig. 7 on the left. This is a byword for splitting the pipe route task into subtasks. Following a degrees of freedom (DoF) based approach, the reference path and the pipe route is split at the breakthrough areas of the route: In a sequentialized solution process tasks with less degrees of freedom have to be conducted first, as presented in [22, 34], to maximize the probability to find a feasible solution. This principle is realized by splitting the pipe route in the breakthrough areas as the degrees of freedom of the pipe route are locally limited in these areas.

Figure 7 shows the result of the pipe route synthesis for a diagonal reference path using the divide and conquer approach. The used installation space is identical to the ship body in the example of Fig. 4. The runtime reduces from hours to minutes when using the divide and conquer approach with the DoF-based task splitting. Additionally, Fig. 7 shows the superiority to existing grid-based methods like [18]: Pipe routes with non axis aligned start/end directions in combination with non orthogonal bend angles.

4 Discussion

The presented algorithm is able to automatically synthesize constrainted pipe routes in given installation spaces. It is limited to the generation of single pipe routes as the rules that change the pipe bends won't preserve the alignment within pipe line corridors with paralelly running pipes. The algorithm is able to synthesize pipe routes made up from standardized pipe bends and flexible length straight elements. Both are available as standardized elements that only have to be cut to the desired lengths in the case of the straight pipe fittings.

Solving this highly constrainted engineering problem takes some hours when treated as a whole. However, applying a divide and conquer scheme by splitting the route into subroutes at pipe routes' positions that are determined by a DoF-based approach leads to a significant reduction of the runtimes by one to two orders of magnitude. On the downside, the divide and conquer approach needs additional effort in kind of the split positions and directions. But this could be potentially automatized within an integrated knowledge-based engineering framework. The runtimes are still higher as for the grid-based methods due to the increase of the search space by the non axis aligned start and end directions in combination with non rectangular pipe bends. The application of the presented method to problems with a high number of pipe lines requires an additional speed up. This could be realized by a parallelization of the synthesizing process in the future.

The presented algorithm has been used in the pipe generation of low lot size exhaust aftertreatment systems that were automatically generated in complex ship or construction machinery powertrain engineering. The algorithm performed well regarding back pressure optimized pipe routes. It was able to generate pipes that could be easily manufactured from standard pipe fittings, on the job, during the installation of the powertrains.

References

1. Voloshin, V.I. (Hrsg.): Introduction to Graph Theory. Published by Nova Science Publishers Inc, New York (2009)
2. Bellman, R.: On a routing problem. Q. Appl. Math. **16**, 87–90 (1958)
3. Dijkstra, E.W.: A note on two problems in connexion with graphs. Numer. Math. **1**, 269–271 (1959)
4. Flloyd, R.W.: Algorithm 97: shortest path. Commun. ACM **5**, S. 345 (1962)
5. Hart, P.E., Nilsson, N.J., Raphael, B.: Correction to: a formal basis for the heuristic determination of minimum cost paths. SIGART Newslett. **37**, 28–29 (1972)
6. Koh, C.-K., Madden, P.H.: Manhattan or non-Manhattan?: a study of alternative VLSI routing architectures. In: Proceedings of the 10th Great Lakes symposium on VLSI. ACM (GLSVLSI), S. 47–52 (2000)
7. Lee, C.Y.: An algorithm for path connections and its applications. In: IRE Transactions on Electronic Computers EC-10, vol. 2, S. 346–365 (1961)
8. Soukup, J.: Global router. In: Proceedings of the 16th Design Automation Conference, pp. 481–484. IEEE Press, Piscataway, NJ, USA (1979) (DAC '79)
9. Ito, D. (Hrsg.): Robot vision: strategies, algorithms and motion planning. Nova Sci. (2009). ISBN 9781606920916
10. Latombe, J.C.: Robot Motion Planning. Springer (1990). (The Springer International Series in Engineering and Computer Science). ISBN 9780792391296
11. Szlapczynski, R.: An algorithm for path connections and its applications. J. Navig. **59**, 27–42 (2006)
12. Velden, C.V., Bill, C., Yu, X., Smith, A.: An intelligent system for automatic layout routing in aerospace design. Innov. Syst. Soft. Eng. **3**, 117–128 (2007)
13. Guirardello, R., Swaney, R.E.: Optimization of process plant layout with pipe routing. Comput. Chem. Eng. **30**, Nr. 1, 99–114 (2005). doi:10.1016/j.compchemeng.2005.08.009. ISSN 0098–1354

14. ITO, T.: A genetic algorithm approach to piping route path planning. In: J. Intell. Manufact. **10**, 103–114 (1999). doi:10.1023/A:1008924832167. ISSN 0956–5515
15. Medjdoub, B.: Constraint-based adaption for complex space configuration in building services. J. Inf. Technol. Constr. 153–158 (2009)
16. Kang, S.-S., Sehyun, M., Han, S.-H.: A design expert system for auto-routing of ship pipes. J. Ship Prod. **15**, 1–9 (1999)
17. Ikehira, S., Kimura, H.: Multi-objective genetic algorithms for pipe arrangement design. In: Proceedings of the Genetic and Evolutionary Computation Conference (GECCO 2006), S. 1869–1870 (2006)
18. Ando, Y., Kimura, H.: An automatic piping algorithm including elbows and bends. In: International Conference on Computer Applications in Shipbuilding, S. 153–158 (2011)
19. Pahl, G. (Hrsg.), Beitz, W. (Hrsg.): Konstruktionslehre, Grundlagen erfolgreicher Produktentwicklung, Methoden und Anwendung. Springer (2003–2005)
20. Norm: DIN EN 10253-2:2008-09, Butt-Welding Pipe Fittings. Beuth Verlag (2008)
21. Norm: DIN 86009:2016-05, Exhaust Gas Lines on Ships—Steel Tubes. Beuth Verlag (2016)
22. Vogel, S.: Über Ordnungsmechanismen im wissensbasierten Entwurf von SCR-Systemen (to appear). Universität Stuttgart, Diss. (2016)
23. Stiny, G.: Shape: Talking About Seeing And Doing. Mit Press (2006) http://books.google.de/books?id=xQpRAAAAMAAJ. ISBN 9780262195317
24. Stiny, G., Gips, J., Stiny, G., Gips, J.: Shape Grammars and the generative specification of painting and sculpture. In: Segmentation of Buildings for 3DGeneralisation, Proceedings of the Workshop on generalisation and multiple representation. Leicester (1971)
25. Antonsson, E., Cagan, J.: Formal Engineering Design Synthesis. Cambridge University Press (2001)
26. Rudolph, S.: Übertragung von Ähnlichkeitsbegriffen. Universität Stuttgart, Habilitationsschrift (2002)
27. Prusinkiewicz, P., Lindenmayer, A.: The Algorithmic Beauty of Plants. Springer (1996). (The Virtual Laboratory). ISBN 9780387946764
28. Szykman, S., Cagan, J.: Synthesis of optimal nonorthogonal routes. In: J. Mech. Des. **118**, Nr. 3, 419–424 (1996). doi:10.1115/1.2826902
29. Kirkpatrick, S., Gelatt, C.D., Vecchi, M.P.: Optimization by simulated annealing. In: Science **220**, 4598 (13 May 1983), 671–680. http://citeseerx.ist.psu.edu/viewdoc/summary?doi=10.1.1.18.4175
30. Featherstone, R.: Rigid Body Dynamics Algorithms. Springer, (2008). (Kluwer international series in engineering and computer science: Robotics). http://books.google.de/books?id=UjWbvqWaf6gC. ISBN 9780387743158
31. Smith, R.: ODE—Open Dynamics Engine. http://www.ode.org. Version: 2007. The Open Dynamics Engine (ODE) is a physics engine in C/C++. Its two main components are a rigid body dynamics and a collision detection
32. Fitzpatrick, R.: Euclid's Elements. Lulu.com, Book 3 (2007)
33. Kröplin, B., Rudolph, S.: Entwurfsgrammatiken—Ein Paradigmenwechsel? Der Prüfingenieur **26**, 34–43 (2005)
34. Vogel, S.: Mathematische Dimension im Entwurf komplexer Systeme, TdSE 2015 (Tag des Systems Engineering) Ulm (2015)

Assessment of Resilience in Desalination Infrastructure Using Semi-Markov Models

Abdulaziz Khiyami, Andrew Owens, Abdelkrim Doufene, Adnan Alsaati and Olivier de Weck

Abstract As the supply of desalinated water becomes significant in many countries, the reliable long-term operation of desalination infrastructure becomes paramount. As it is not realistic to build desalination systems with components that never fail, instead the system should be designed with more resilience. To answer the question how resilient the system should be, we present in this paper a quantitative approach to measure system resilience using semi-Markov models. This approach allows to probabilistically represent the resilience of a desalination system, considering the functional or failed states of its components, as well as the probability of failure and repair rates. As the desalination plants are connected with the end-user through water transportation and distribution networks, this approach also enables an evaluation of various network configurations and resilience strategies. A case study addressing a segment of the water system in Saudi Arabia is given with the results, benefits, and limitations of the technique discussed.

Keywords System resilience • Water system • Semi-Markov process

List of Acronyms

CDF	Cumulative Distribution Function
MRP	Markov Renewal Process
MTBF	Mean Time Between Failures
MTTR	Mean Time to Repair
PDF	Probability Distribution Function
SDR	Standard Deviation in Repair Time

A. Khiyami (✉) · A. Alsaati
Center for Complex Engineering Systems at KACST and MIT,
PO Box 6086, Riyadh 11442, Kingdom of Saudi Arabia
e-mail: a.khiyami@cces-kacst-mit.org

A. Owens · A. Doufene · O. de Weck
Massachusetts Institute of Technology, 77 Massachusetts Avenue,
Building E40-261, Cambridge, MA 02139-4307, USA
e-mail: acowens@mit.edu

G. Fanmuy et al. (eds.), *Complex Systems Design & Management*,
DOI 10.1007/978-3-319-49103-5_10

SMP Semi-Markov Processes
SWCC Saline Water Conversion Corporation

1 Introduction

Water is a prerequisite for life and its provision in modern society is contingent on numerous interacting components that include the water source; physical infrastructure; the services it provides; the organizations that govern its use; and the people and industry that consume it, and produce waste water. As the interdependence between these components is strong, and in order to make water use more efficient, together these components may collectively be aggregated in one system, that we call the 'water system' in this paper.

Given water's criticality, water system planners must continuously assess and manage a host of challenges to ensure the satisfactory performance of their systems. These challenges include the ever-present need to balance costs and impacts to the environments as well as the preparation for a variety of potential hazards such as natural disasters, and terrorist attacks, etc. This undertaking requires a continuous cycle of evaluation and planning activities following adverse events to upgrade and adapt the water system based on lessons learned. In an effort to aid and quantify this process, numerous attributes and objectives with which to assess the performance of water systems have been proposed. These include but are not limited to: cost, sustainability, reliability, robustness, preparedness, responsiveness, vulnerability, etc.

Key among these many overlapping and oftentimes conflicting objectives has been the concept of water system 'resilience'. Resilient systems have been described in the literature as those with "the ability to reduce the magnitude and/or the duration of disruptive events" [1] or "the ability to minimize the costs of a disaster, return to the status quo, and to do so in the shortest feasible time" [2, 3] define resilience as "the capacity for a system to survive, adapt, and flourish in the face of turbulent change and uncertainty." Hashimoto [4] describes resilience as one of three key special risk-related system performance criteria in the widely utilized Reliability, Resiliency, and Vulnerability (RRV) framework and defines it as "how quickly a system is likely to recover or bounce back from failure once failure has occurred."

Figure 1 graphically illustrates these definitions of system resilience. The function F(t) may represent any system performance measure provided that higher values correlate to higher performance. At a time T_e, the systems performance has fallen below a prescribed failure threshold entering a Disrupted State. Following a resilience action to repair the system, performance reaches above the failure threshold at time T_r. The difference $T_r - T_e$ is the time spent in a failed (disrupted) state. The design of a resilient system should seek to minimize this time period, crafting systems that are both unlikely to fall below the prescribed failure threshold and quickly recover from failure should a failure occur.

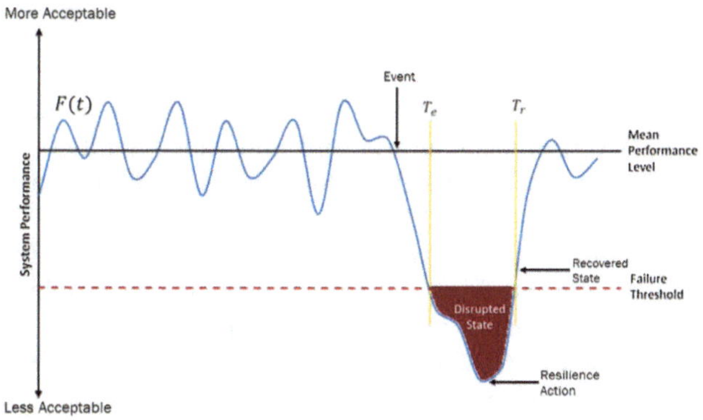

Fig. 1 Graph of Resilience. Adapted from Hashimoto [4]

A review of the literature on resilience reveals that many of its aspects bear similarity to the concepts of risk, reliability, preparedness, vulnerability assessment, disaster management and risk management. The question thus becomes: how does resilience differ from these concepts; and is it a distinct concept or just a different word for the same activities?

Resilience is indeed heavily intertwined with these concepts, however there appears to be a consensus that its key lies in the anticipation of unexpected events [5].

In this vein we propose a probabilistic framework devised using semi-Markov models to quantitatively model and assess the expected resilience of a water system. Each component in the system is defined by its status (functional/failed) and transition probability distributions defined by failure rates, repair rates, and the time that the system can maintain its performance after component failures. This technique enables the calculation of all likely potential system states, and the probability of system failure within a chosen study period, thereby anticipating conceivable system failures.

To exemplify this approach we analyze a case study from the Kingdom of Saudi Arabia, an arid country that has turned to desalination for much of its municipal water supply. With a heavy reliance upon desalination and an extensive network of plants and pipelines, the Kingdoms water system performance is especially beholden to plant outages, pipe breaks, and pump failures. These failure conditions are easily anticipated but occur unexpectedly. Our approach provides a framework for these events to be anticipated and planned for so that they are less disruptive to the overall system performance, thereby increasing resilience.

The paper is organized as follows: in the Background section we present the context of Saudi Arabia, in the Methodology section we discuss the theoretical and mathematical procedures of the resilience framework; and in the Application section we utilize the developed methodology for the Saudi context. Future work and conclusions are presented in the final section.

2 Background

The Kingdom of Saudi Arabia (KSA) is the largest country in the world with no permanent natural rivers or lakes, an arid land with seldom rainfall. As such the vast majority of water consumed in the Kingdom comes from non-renewable ground water resources (SSDN [6]; SIPS [7]).

To compensate for its lack of natural freshwater the Kingdom has increasingly turned to desalination to satisfy its water needs. Today Saudi Arabia is the world's largest market for desalinated water with a capacity of 5.72 million m^3/day accounting for as much as 60 % of the total urban water supply [8].

Perhaps no city can better demonstrate Saudi Arabia's extreme reliance upon desalination better than its capital Riyadh. Initially a small oasis town of no more than 10,000 inhabitants [9] at the start of the 20th century the capital is now a bustling metropolis with a population close to 7 million. Having long ago outgrown its local water resources, Riyadh now meets nearly half its municipal demand from desalinated water that is produced at giant facilities on the East Coast and then pumped via pipelines over hundreds of kilometers.

Thus, the optimal operation of the desalination system depends not only upon the stand-alone plants but the network as a whole. The evaluation of a desalination system as network of production nodes (desalination plants) and consumption nodes (cities) connected by edges of water pipelines is therefore informative for enhancing the design of the system in its entirety.

Ishimatsu et al. [10] presented such a deterministic network model that allowed for a desalination network's optimization in space, that is, where geographically a new infrastructure component should be located at a given time. This procedure utilized a graph theoretic framework with a multi-objective optimization to design the network for cost and/or sustainability.

However, the stated mission of the Saline Water Conversion Corporation (SWCC), the main institution tasked with the supply of desalinated water, is *the secure and maintained provision of water to the nation.*

Therefore to truly optimize KSA's desalination network, a model that considers failure and resilience is necessary. An optimization that only considers nominal operating conditions is not realistic indeed as it will overestimate the systems capabilities and underestimate its operating costs.

3 Methodology

In this paper, we utilize Semi-Markov Processes (SMPs) to examine the resilience of water pipeline networks for a given operating duration, looking in particular at the amount of downtime, the amount of unmet demand, and the number of repair actions that will be required. All of these metrics are stochastic, not deterministic, since the underlying processes behind them—failures and repairs—are inherently stochastic. As such, the outputs of the model are not single point values, but rather distributions. These can then be used by decision-makers to make risk-informed decisions regarding local storage capacity, resource allocation for maintenance actions, and operating cost projections.

Semi-Markov Processes

SMPs are probabilistic, state-based models of system behavior that are an extension of Markov chains. Like Markov chains, SMPs represent system behavior in a directed graph of states and transitions, where states (nodes) represent a given configuration of the system and transitions (edges) are events that cause the system configuration to change from one state to another. Each transition has an associated probability distribution which describes the amount of time until that transition occurs once the state it leaves is entered. An important requirement on SMPs is that, similar to Markov chains, the states must be "memoryless," meaning that the future evolution of the system is dependent only on the current state and not on the pathway taken to reach that state. However, whereas in Markov chains these distributions must be exponential, SMPs allow the use of any distribution [11–13]. An excellent overview of SMPs and techniques for solving them is presented by Warr and Collins [13].

An SMP is fully characterized by the kernel matrix $Q(t)$ and the unconditional waiting time density matrix $H(t)$, each of which have entries that are calculated as follows (Warr and Collins [13]:

$$Q_{ij}(t) = f_{ij}(t) \prod_{k \neq j} \left(1 - \int_0^t f_{ij}(\xi) d\xi \right)$$

$$H_{ii}(t) = \sum_j Q_{ij}(t)$$

where $f_{ij}(t)$ is the Probability Distribution Function (PDF) describing the amount of time t that passes after entry into state i before a transition from state i to state j occurs, given that a transition to state j does occur (as opposed to some other state). Each entry $Q_{ij}(t)$ of the kernel matrix is a PDF describing the amount of time t that passes after entry into state i before a transition from state i to state j occurs, assuming no transition to any other state occurs in the interim. This can be seen from the fact that it is a product of the PDF of the time until transition from state i to state j and the complements of the Cumulative Distribution Functions (CDFs) of all other transitions. The unconditional waiting time density matrix is a diagonal matrix with entries $H_{ii}(t)$ that give the PDFs describing the amount of time t that passes after entry into state i until a transition out of state i occurs, regardless of the destination state. Given $Q(t)$ and $H(t)$, several key metrics describing the behavior of the system modeled by the SMP can be solved for. These metrics are listed in

Table 1 [13]. The process of calculating these metrics from $Q(t)$ and $H(t)$ using the Laplace domain is described in greater detail below.

Application to Resilience Modeling

SMPs have previously been used to examine the resilience and maintenance logistics requirements of space systems [14–20], and we use a similar approach here. In this formulation, each state in the SMP is characterized in terms of the status—functional or failed—of each element—pipeline or desalination plant—within the system.

As is suggested by the state formulation, the transitions between states represent failure and repair events. (In the case where degraded states are included, these would include degradation and partial repair events.) The PDF used depends on the transition being represented. Failures are characterized by exponential distributions – a common first-order model of random component failures known as the constant failure rate model [21]. The rate parameter of this distribution is equal to the inverse of the Mean Time Between Failures (MTBF) for each particular element. Repairs are modeled using a lognormal distribution, which provides a good estimate of the time required for corrective repair [22, 23]. In this case, the distribution is formed to have a mean and standard deviation equal to the Mean Time to Repair (MTTR) and Standard Deviation in Repair Time (SDR) for each particular repair activity.

The structure of the network of states and transitions representing the SMP is specifically constructed to link the generic SMP metrics described in Table 1 to system metrics. In particular, the structure of the SMP links the Markov Renewal Process (MRP) probabilities—which give the distribution of the number of times a given state will be visited in a given period of time—to the number of failures experienced by a particular element by ensuring that each state is linked to the failure of a particular component. This is done by ensuring that every state is

Table 1 Symbols, names, and descriptions of key SMP metrics. All metrics assume that the system starts in state i at time 0 [13]

Symbol	Name	Description
$\phi_{ij}(t)$	Time-dependent state probability	Probability that the system will be in state j at time t
$E_{ij}(t)$	Expected time in state	Expected amount of time that the system will have spent in state j up to time t
$g_{ij}(t)$	PDF of first passage time	PDF describing the time t taken to reach state j the first time
$G_{ij}(t)$	CDF of first passage time	CDF giving the probability that the system has reached state j by time t
$V_{ij}(k, t)$	MRP probability	CDF giving the probability that the system has reached state j a total of k or fewer times by time t

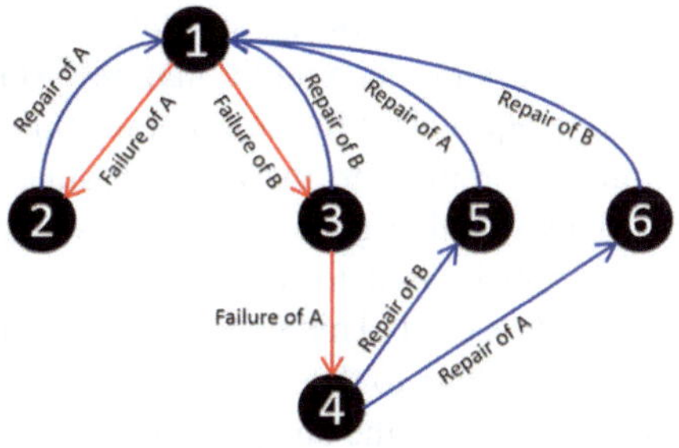

Fig. 2 Example SMP state/transition network for a system with two elements, A and B. Each transition is labeled with the event it represents. *Red* transitions indicate failure events, and *blue* transitions indicate repairs

entered by one and only one failure transition. Therefore, the number of times that a given state is visited corresponds to the number of times that that failure occurs. An example of this network structure is given in Fig. 2. When multiple states are entered by failure of the same element, the MRP distributions for these states are convolved together to determine the total number of failures experienced by that element. Additional details on the connection between state structure and system metrics, as well as restrictions on SMP structure, are discussed by Owens [16].

The impacts of failures are captured via the state definitions. Since each state is characterized by the status of each element within the system, a model that can characterize system performance as a function of element status can then produce key metrics for each state, such as the rate of unmet demand in a given city. This information can be combined with SMP metrics relating to states, including distributions for the number of times a state is visited and the amount of time spent in that state, in order to develop distributions for these key metrics [17].

Automated SMP Generation

A key limitation for the application of SMPs to systems analysis of this type is that the number of states that a given system could be extremely large. As a result, the generation of the SMP model itself can be a very time-consuming process unless some form of automation can be utilized. While some previous applications of SMPs have used manually-generated state network models that limit state-space with simplifying assumptions [16, 19, 20], we implement an automated SMP generation algorithm based on one presented previously for space systems by Owens and de Weck [17].

The algorithm consists of a systematic enumeration of new states based on existing ones, starting from the nominal state (i.e. all systems operational). New states—called "children" of the current state—are produced by examining all possible transitions away from the current state. In general, elements that are currently functional can fail, and elements that are currently failed can be repaired. For example, the nominal state has a set of transitions away from it representing the failure of each element in the system, each of which ends at a new state representing the configuration of the system in which that element is failed. Additional failures and repairs produce additional new states, unless the configuration of the resulting state is equivalent to the nominal state (all systems operational), in which case the transition returns to the nominal state rather than creating a new state [17].

This iterative generation of new states grows the SMP network, and a pruning algorithm is used to remove states that have a probability of occurrence below a given threshold. This is done by calculating the first passage probability $G_{ij}(t)$ for each new state to determine the probability that it is visited at least once within the time horizon of the analysis; if this probability is below a given threshold, and if the state was entered by a failure event and not a repair event, the state is removed from the network. States entered by repair events are not removed from the network since they are a part of the pathway back to the nominal state, forming the loops that enable the use of MRP probabilities to examine spares requirements [17].

The main difference between the algorithm used here and the one described by Owens and de Weck [17] is that in this case new states are produced in generations, rather than one at a time, before pruning is applied. Generation 0 is the nominal state, generation 1 consists of all of the children of the nominal state, generation 2 consists of all the children of the children of the nominal state, and so on. Pruning of states in generations rather than individually significantly decreases the amount of computational time required to generate the SMP network.

Model Solution

Once an SMP model of the system is produced, it can be solved for the key metrics of interest. This process consists of two steps. First, the SMP is solved for the metrics described in Table 1, or whatever subset of them is desired for a particular problem. In this case, we are particularly interested in the MRP probabilities $V_{ij}(t)$, which are partially based on the first passage time PDFs $g_{ij}(t)$. These metrics can be solved for quickly using matrix multiplication in the Laplace domain followed by numerical Laplace transform inversion [13]. For convenience, following the convention of Warr and Collins [13], we abbreviate the symbol for the Laplace transform as a tilde (~) over the relevant matrix. The equations for first passage time and MRP probabilities in the Laplace domain are:

$$\tilde{g}(s) = \tilde{Q}(s)\left(I - \tilde{Q}(s)\right)^{-1}\left(I \circ \left(I - \tilde{Q}(s)\right)^{-1}\right)^{-1}$$

$$\tilde{V}(k, s) = \frac{1}{s}\left(1 - \tilde{g}(s) \circ \left(1 (I \circ \tilde{g}(s))^{k}\right)\right)$$

where I is the identity matrix, $\circ$. The Hadamard product of two matrices (elementwise multiplication), and 1 is a matrix of ones [13]. Once the Laplace transform of the MRP probabilities is obtained using the equations above, the EULER numerical Laplace transform inversion technique developed by Abate and Whitt [24] is utilized to obtain the time-domain MRP probabilities. Owens [16] presents a brief overview and explanation of the numerical Laplace and inverse Laplace transform algorithms used here in Appendix A of his thesis, and more detail, including derivations and background, is presented by Warr and Collins [13] and Abate and Whitt [24].

The result of the above procedure is the distribution of the number of times each state in the SMP is visited. This result can be used directly to determine the distribution of the number of failures that each element in the system will experience, as described above. When combined with the unconditional waiting time density $H_{jj}(t)$ for each state j, the distribution of the number of visits to state j (assuming a start in state 0, the nominal state) $V_{0j}(k,t)$ can also be used to generate $T_j(t)$, the distribution of the total amount of time that will be spent in state j for the time period examined.

$$T_j(t) = V_{0j}(0,t)(\delta(0)) + \sum_{k=1}^{\infty} \left(V_{0j}(k,t) - V_{0j}(k-1,t)\right)\left(\mathrm{Conv}_k\left(H_{jj}(t)\right)\right)$$

Here $\delta(0)$ is the Dirac delta function and $\mathrm{Conv}_k(f(t))$ is a function representing the convolution of k instances of a function $f(t)$—that is, $\mathrm{Conv}_1(f(t)) = f(t)$, $\mathrm{Conv}_2(f(t)) = f(t)*f(t)$, and so on. When applied to the unconditional waiting time density for a particular state, this convolution produces the distribution of the total amount of time spent in that state given that the state is visited exactly k times. This distribution is then conditioned by the probability that the state is visited exactly k times, and the sum of these conditioned distributions (representing the possible cases for the number of times the state will be visited) gives the distribution of the total amount of time spent in that state. In practice, the summation in the equation above is only carried out as far as there is a non-negligible probability of k visits to the state rather than continuing to infinity.

As described above, each state in the SMP is characterized by the status of each element within it. For this case study, this means the status of each pipeline and desalination plant as either functional or failed. For high-level decision-making, however, a more relevant metric of interest may be the impact of these failures on water delivery to consumers (in this case, cities). Therefore, each state is characterized in terms of the rate of unmet demand at each city by solving an optimization problem to determine the flow configuration in the network that minimizes the total rate of unmet demand across all cities. In the nominal state, each pipeline and desalination plant has a maximum capacity indicating the amount of water it can transport or produce. States in which a failure has occurred in one or more elements have the capacities of that element set to zero in order to simulate the impacts of that failure. This reduction in network capability results in reduced ability to meet consumer demands, which in turn results in some rate of unmet demand at some (or all) of the cities in the network. The optimization problem for a system with n cities and m elements (pipelines and desalination plants) is formulated as follows:

$$
\begin{aligned}
&\text{minimize} \quad \sum_{i=1}^{n} u_i \\
&\text{subject to: } u_i + \sum_{j \in \mathrm{IN}_i} x_j - \sum_{j \in \mathrm{OUT}_i} x_j = d_i \quad \forall i \in \{1, \ldots, n\} \\
&0 \le u_i \le d_i \quad \forall i \in \{1, \ldots, n\} \\
&-c_j \le x_j \le c_j \quad \forall j \in \{1, \ldots, m\}
\end{aligned}
$$

where u_i is the rate of unmet demand at city i, d_i is the rate of demand at city i, c_j is the flow capacity for element j, x_j is the flow rate in element j, and IN_i and OUT_i are the sets of elements flowing into and out of city i, respectively. Note that self-loops, which represent desalination plants, appear only in the set of elements flowing into their city, and not the set flowing out. This linear optimization problem is quickly and easily solved using MATLAB's built-in linprog() function in order to determine the rate of unmet demand at each city in each state of the SMP.

It is possible that some states in the SMP are identical in terms of their system-level characteristics. Therefore, once the amount of time spent in each state and the rate of unmet demand for each city in each state are determined, the distributions for the amount of time spent in states with identical unmet demand profiles are convolved together to determine the total amount of time the system spends in that condition. Alternatively, these distributions could be convolved together based on the unmet demand rate for a particular city. Once the distribution of the total amount of time spent at a given rate of unmet demand is obtained, it can be used with the specific rate of unmet demand to determine the distribution of the total amount of unmet demand in the time period being examined, which can then be used to inform storage capacity decisions.

4 Application

The proposed methodology is applied to a subsection of Saudi Arabia's easterly desalination network. Figure 3 (left) shows the system containing the capital city of Riyadh and associated desalination plants and cities on the Arabian Gulf. Though in reality the network extends beyond Riyadh, and also branches out onto other Eastern cities, for this case study the analysis is focused upon the largest and most significant population centers of the region The simplified network representation considered in the case study is shown in Fig. 3 (right).

Network Case Study Parameters

The parameters of the desalination network are recorded in Tables 2 and 3 with the chosen analysis units of cubic meters and days. Daily city desalinated water

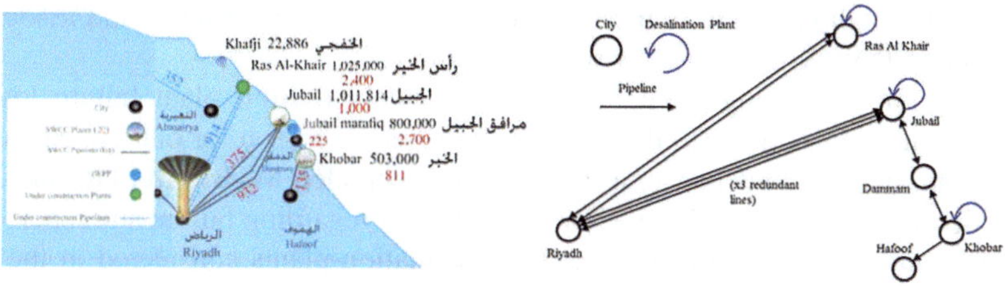

Fig. 3 Eastern desalination network [29] and case study representation

Table 2 Node parameters

Node ID	Node name	Demands (1000 m^3/day)
1	Riyadh	701
2	Ras Al Khair	0
3	Jubail	42
4	Dammam	113
5	Khobar	572
6	Hafoof	83

Table 3 Edge parameters

ID from	ID to	Name	MTBF (days)	MTTR (days)	SDR (days)	Edge capacities (1000 m^3/day)
2	2	Ras Al Khair Desalination Plant	60	4	3	1025
3	3	Jubail Desalination Plant	60	4	3	1782
5	5	Khobar Desalination Plant	60	4	3	547
1	2	Riyadh—Ras Al Khair D	110	14	7	474
1	2	Riyadh—Ras Al Khair E	110	14	7	474
1	3	Riyadh—Jubail A	110	14	6	415
1	3	Riyad Jubail B	100	14	6	415
1	3	Riyadh—Jubail C	100	14	6	380
3	4	Jubail—Dammam	90	5	1	305
4	5	Dammam—Khobar	75	4	1	305
5	6	Khobar—Hafoof	80	5	1	266

demands were calculated using the population, per capita daily water consumption, and percentage contribution of desalination in a manner similar to the methodology previously utilized by Ishimatsu et al. [10]. Desalination plant capacities and pipeline throughputs were found as specified in designs by SWCC and associated contractors (SWCC [25] and Lasser and Heinz [26]).

Indications regarding plant failures were received from plant failure logs of SWCC. These logs included the duration and specific reason for outages e.g. steam line leaks, boiler maintenance; as well as the calculated MTBF, MTTR, and SDR for a desalination plant in 2015. Exact information regarding failure and repair rates was not made available for the specific desalination plants considered in the case study, and so the provided plants MTBF and MTTR were used as representative.

Information on failure and repair rates of pipelines was not forthcoming and was therefore estimated from news reports [27], technical reports [28], and the recommendations of SWCC staff. On average, desalination pipelines were found to break less often than desalination plants, but require longer to repair.

The data therefore used in this case study is merely notional and intended to only demonstrate the proposed methodology, not to provide concrete results or recommendations.

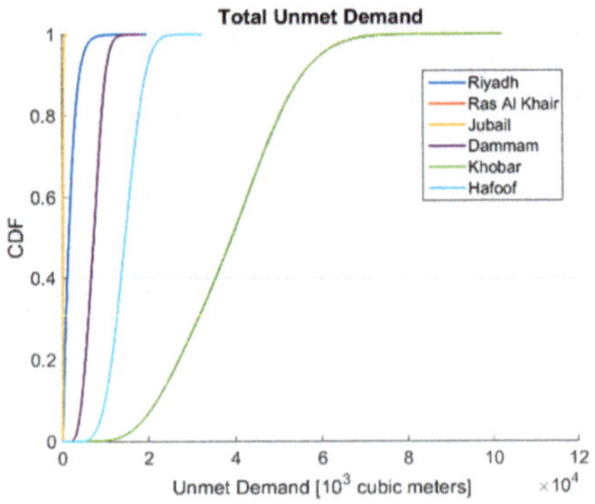

Fig. 4 CDF of Total Unmet Demand

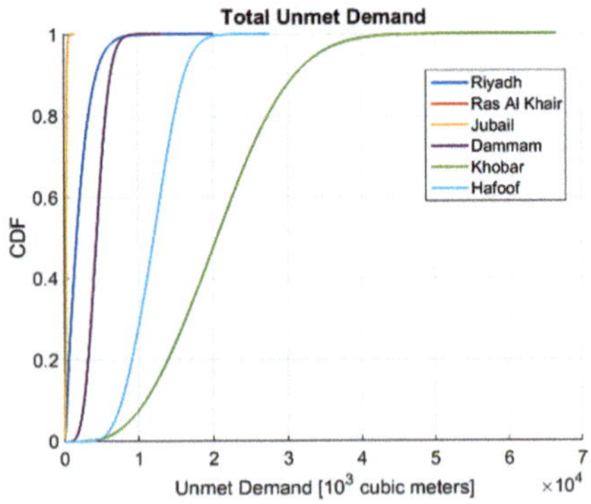

Fig. 5 CDF of Total Unmet Demand for Khobar MTBF 120

Case Study Execution and Results

The model was formulated in MATLAB and executed for a time horizon of 10 years with a state probability threshold of 0.25 %. Computationally this required about 15 min of running time on a single machine using an Intel® Xeon® CPU E5-2650 v3 with 32 GB of installed RAM.

The CDF of unmet demand for each city was calculated and this is plotted in Fig. 4. The analysis reveals for example that Riyadh, with its numerous feeder pipelines is relatively safe to the risk of unmet demand with nearly an 80 % probability that unmet demand will not exceed 3 million m^3 throughout the 10 years considered. Strategic reserves of only 1 million m^3 are necessary to ensure that the city has a near zero chance of any unmet demand.

By contrast the Eastern Region cities of Dammam, Khobar, and Hafoof are far more vulnerable with Khobar, the largest of the three, being most at risk. Throughout the same 10 year period, Khobar has an 80 % probability of experiencing nearly 50 million m^3 of unmet demand and would require reserves of 75 million m^3 to ensure against failure. This is intuitive, Khobar approaches Riyadh in its daily desalination demand but does not have the benefit of a direct connection to the Ras Al Khair facility or anywhere near as many redundant feeder pipelines.

To design for system resilience various strategies can now be explored using the proposed approach. For example adding a new desalination plant at Dammam, or connecting Ras Al Khair to Jubail with a new pipeline. Increasing plant/pipeline reliability through upgrades and more vigilant maintenance of the network elements can be investigated via variance of the failure and repair rates.

It was discovered that among the most effective ways to reduce the risk of unmet demand was by improving the reliability of the Khobar desalination plant. Doubling the MTBF from once every 60 days to once every 120 days reduces the expected unmet demand at probability of 80 % by nearly half as shown in Fig. 5. Further increasing the reliability of the Khobar desalination plant found further reductions in expected unmet demand but at diminishing returns as shown in Fig. 6.

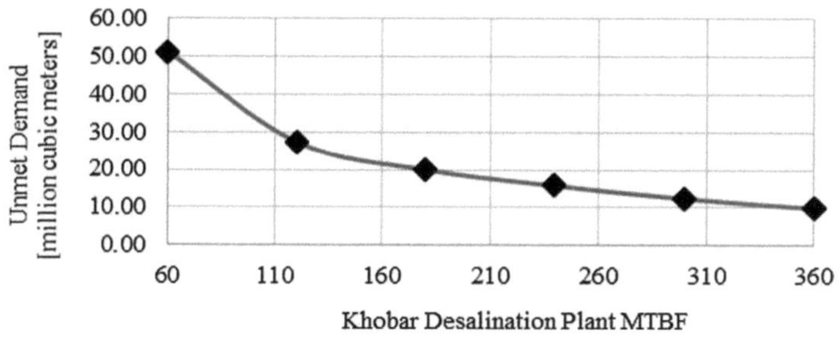

Fig. 6 Khobar 80th percentile unmet demand

5 Conclusions and Further Work

This paper introduces an approach to quantitatively evaluate the resilience of water systems. The modelling procedure was illustrated via a notional case study of a portion of Saudi Arabia's desalination network.

The current approach provides a starting framework upon which to improve for an advanced assessment of resilience in water systems. For starters, the current approach employs a binary fail/repair status for each network element; further work should explore the representation of partially degraded states to more fully represent the operation of the system. The current application utilizes static network demands to evaluate resilience well into the future. A model that incorporates dynamically changing demand and future growth scenarios will contribute to the understanding of how efficiency and end-user programs may affect the system resilience. Additionally the characterization of specific outages and failures needs to be introduced to the framework. For example, if an extreme event could cause all desalination plants to be shut-down simultaneously, the likelihood and consequences of such an event is not currently considered in the model. Finally, the methodology should be enhanced by the implementation of a resilience optimization that will automatically find the best combination of network upgrades and expansions to maximize resilience. Future work should also more holistically evaluate the water system, considering agricultural demands and groundwater reserves, as well as waste water treatment, rather than just the desalination system in isolation to assess the resilience of the water system in its entirety.

Acknowledgments The authors would like to thank all the SSDN project team members for their collaboration, comments and feedback. This research was supported by the Center for Complex Engineering Systems (CCES) at MIT and KACST (http://www.cces-kacst-mit.org/). This research was supported by a NASA Space Technology Research Fellowship (grant number NNX14AM42H).

References

1. NIAC: Critical Infrastructure Resilience Final Report and Recommendations. s.l.: National Infrastructure Advisory Council (2009)
2. Fiksel, J., Goodman, I., Hecht, A.: Resilience: navigating a sustainable future. Solut. Sustain. Desirable Future **5**(5), 38–47 (2014)
3. McAllister, T.: Developing Guidelines and Standards for Disaster Resilience of the Built Environment: a Research Needs Assessment. NIST, Gaithersburg (2013)
4. Hashimoto, T.: Reliability, resiliency, and vulnerability criteria for water resource system performance evaluation. Water Resour. Res. **18**(1), 14–20 (1982)
5. EPA: Systems Measures of Water Distribution System Resilience. United States Environmental Protection Agency, Cincinnati (2015)

6. SSDN—Strategic Sustainable Desalination Network. Final Report, Center for Complex (2015)
7. SIPS—Sustainable Infrastructure Planning System Project: Final Report, Center for Complex Engineering Systems at KACST and MIT (2015). http://www.cces-kacst-mit.org
8. SWCC: Annual Report. Saline Water Conversion Corporation, Riyadh (2014)
9. Al-Naim, M.: Riyadh: a city of 'institutional' architecture. In: Elsheshtawy, Y. (ed.) The Evolving Arab City: Tradition, Modernity and Urban Development, pp. 118–152. Routledge, Abingdon (2008)
10. Ishimatsu, T., Doufene, A., de Weck, O.: Desalination network model driven decision support system: a case study of San Diego. In: International Desalination Association World Congress 2015 (2015)
11. Lisnianski, A., Levitin, G.: Multi-State System Reliability. World Scientific, Singapore (2003)
12. Nunn, W.R., Desiderio, A.M.: Semi-markov processes: an introduction, Arlington, Virginia (1977)
13. Warr, R.L., Collins, D.H.: An introduction to solving for quantities of interest in finite-state semi-Markov processes (2012)
14. Do, S., Owens, A.C., de Weck, O.L.: HabNet—an integrated habitation and supportability architecting and analysis tool. In: 45th International Conference on Environmental Systems, ICES-2015-289, Bellevue, WA (2015)
15. Do, S., Owens, A., Ho, K., Schreiner, S., de Weck, O.: An independent assessment of the technical feasibility of the mars one mission plan—updated analysis. Acta Astronaut. **120**, 192–228 (2016)
16. Owens, A.C.: Quantitative probabilistic modeling of environmental control and life support system resilience for long-duration human spaceflight. Massachusetts Institute of Technology (2014)
17. Owens, A.C., de Weck, O.L.: Automated risk and supportability model generation for repairable systems. In: 66th International Astronautical Congress, IAC-15-D1.3.10, pp. 1–9. International Astronatical Federation, Jerusalem (2015)
18. Owens, A.C., de Weck, O.L.: Use of semi-Markov models for quantitative ECLSS reliability analysis: spares and buffer sizing. In: 44th International Conference on Environmental Systems, ICES-2014-116, Tucson, AZ (2014)
19. Owens, A., de Weck, O., Mattfeld, B., Stromgren, C., Cirillo, W.: Comparison of spares logistics analysis techniques for long duration human spaceflight. In: 45th International Conference on Environmental Systems, ICES-2015-288, Bellevue, WA: International Conference on Environmental Systems (2015)
20. Owens, A., Do, S., Kurtz, A., de Weck, O.: Benefits of additive manufacturing for human exploration of mars. In: 45th International Conference on Environmental Systems, ICES-2015-287, Bellevue, WA: International Conference on Environmental Systems (2015)
21. Ebeling, C.E.: An Introduction to Reliability and Maintainability Engineering. Tata McGraw-Hill Publishing Co., Ltd., New Delhi (2000)
22. Jones, H.: Life support dependability for distant space missions. In: 40th International Conference on Environmental Systems, AIAA-2010-6287. American Institute of Aeronautics and Astronautics, Barcelona (2010)
23. Kline, M.B.: Suitability of the lognormal distribution for corrective maintenance repair times. Reliab. Eng. **9**, 65–80 (1984)
24. Abate, J., Whitt, W.: Numerical inversion of Laplace transforms of probability distributions. ORSA J. Comput. **7** (1995)
25. SWCC: SWCC Projects Pipeline (2016). http://www.swcc.gov.sa/english/Projects/Pipelines/Pages/PipeLine.aspx. Accessed 11 April 2016
26. Lasser, B., Heinz, A.: Innovations and technological developments in long distance water transmission systems: case study of Riyadh, Saudi Arabia. Essen: 3R International Technical Journal for Piping System Integrity and Efficiency (2011)

27. Khan, S.A.: Riyadh residents breathe sigh of relief as water supply restored. Saudi Gazette **16**, 09 (2011)
28. Malik, A.U., Andijani, I., Mobin, M., Al-Hajri, M.: Investigations on the Leakage in RWTS, Line—A at 24 km (A/B)1. SWCC, Al Jubail (2005)
29. SWCC: Annual Report. Saline Water Conversion Corporation, Riyadh (2013)

A Discrepancy-Based Framework to Compare Robustness Between Multi-attribute Evaluations

Juste Raimbault

Abstract Multi-objective evaluation is a necessary aspect when managing complex systems, as the intrinsic complexity of a system is generally closely linked to the potential number of optimization objectives. However, an evaluation makes no sense without its robustness being given (in the sense of its reliability). Statistical robustness computation methods are highly dependent of underlying statistical models. We propose a formulation of a model-independent framework in the case of integrated aggregated indicators (multi-attribute evaluation), that allows to define a relative measure of robustness taking into account data structure and indicator values. We implement and apply it to a synthetic case of urban systems based on Paris districts geography, and to real data for evaluation of income segregation for Greater Paris metropolitan area. First numerical results show the potentialities of this new method. Furthermore, its relative independence to system type and model may position it as an alternative to classical statistical robustness methods.

Keywords Multi-attribute evaluation · Model-independent robustness · Urban system · Discrepancy

1 Introduction

1.1 General Context

Multi-objective problems are organically linked to the complexity of underlying systems. Indeed, either in the field of *Complex Industrial Systems*, in the sense of engineered systems, where construction of Systems of Systems (SoS) by coupling

---

J. Raimbault (✉)
UMR CNRS 8504 Géographie-Cités, Paris, France
e-mail: juste.raimbault@polytechnique.edu

J. Raimbault
UMR-T IFSTTAR 9403 LVMT, Champs-sur-marne, France

G. Fanmuy et al. (eds.), *Complex Systems Design & Management*,
DOI 10.1007/978-3-319-49103-5_11

and integration often leads to contradictory objectives [1], or in the field of *Natural Complex Systems*, in the sense of non engineered physical, biological or social systems that exhibit emergence and self-organization properties, where objectives can e.g. be the result of heterogeneous interacting agents (see [2] for a large survey of systems concerned by this approach), multi-objective optimization can be explicitly introduced to study or design the system but is often already implicitly ruling the internal mechanisms of the system. The case of socio-technical Complex Systems is particularly interesting as, following [3], they can be seen as hybrid systems embedding social agents into "technical artifacts" (sometimes to an unexpected degree creating what PICON describes as *cyborgs* [4]), and thus cumulate propensity to be at the origin of multi-objective issues.[1] The new notion of *eco-districts* [5] is a typical example where sustainability implies contradictory objectives. The example of transportation systems, which conception shifted during the second half of the 20th century from cost-benefit analysis to multi-criteria decision-making, is also typical of such systems [6]. Geographical system are now well studied from such a point of view in particular thanks to the integration of multi-objective frameworks within Geographical Information Systems [7]. As for the micro-case of eco-districts, meso and macro urban planning and design may be made sustainable through indicators evaluation [8].

A crucial aspect of an evaluation is a certain notion of its reliability, that we call here *robustness*. Statistics naturally include this notion since the construction and estimation of statistical models give diverse indicators of the consistence of results [9]. The first example that comes to mind is the application of the law of large numbers to obtain the *p-value* of a model fit, that can be interpreted as a confidence measure of estimates. Besides, confidence intervals and *beta-power* are other important indicators of statistical robustness. Bayesian inference provide also measures of robustness when distribution of parameters are sequentially estimated. Concerning multi-objective optimization, in particular through heuristic algorithms (for example genetic algorithms, or operational research solvers), the notion of robustness of a solution concerns more the stability of the solution on the phase space of the corresponding dynamical system. Recent progresses have been done towards unified formulation of robustness for a multi-objective optimization problem, such as [10] where robust Pareto-front as defined as solutions that are insensitive to small perturbations. In [11], the notion of degree of robustness is introduced, formalized as a sort of continuity of other solutions in successive neighborhood of a solution.

However, there still lack generic methods to estimate robustness of an evaluation that would be model-independent, i.e. that would be extracted from data structure and indicators but that would not depend on the method used. Some advantages could be

[1]We design by *Multi-Objective Evaluation* all practices including the computation of multiple indicators of a system (it can be multi-objective optimization for system design, multi-objective evaluation of an existing system, multi-attribute evaluation; our particular framework corresponds to the last case).

for example an *a priori* estimation of potential robustness of an evaluation and thus to decide if the evaluation is worth doing. We propose here a framework answering this issue in the particular case of Multi-attribute evaluations, i.e. when the problem is made unidimensional by objectives aggregation. It is data-driven and not model-driven in the sense that robustness estimation does not depend on how indicators are computed, as soon as they respect some assumptions that will be detailed in the following.

1.2 Proposed Approach

Objectives as Spatial Integrals We assume that objectives can be expressed as spatial integrals, so it should apply to any territorial system and our application cases are urban systems. It is not that restrictive in terms of possible indicators if one uses suitable variables and integrated kernels: in a way analog to the method of geographically weighted regression [12], any spatial variable can be integrated against regular kernels of variable size and the result will be a spatial aggregation which sense depends on kernel size. The example we use in the following such as conditional means or sums suit well the assumption. Even an already spatially aggregated indicator can be interpreted as a spatial indicator by using a Dirac distribution on the centroid of the corresponding area.

Linearly Aggregated Objectives A second assumption we make is that the multi-objective evaluation is done through linear aggregation of objectives, i.e. that we are tackling a multi-attribute optimization problem. If $(q_i(\vec{x}))_i$ are values of objectives functions, then weights $(w_i)_i$ are defined in order to build the aggregated decision-making function $q(\vec{x}) = \sum_i w_i q_i(\vec{x})$, which value determines then the performance of the solution. It is analog to aggregated utility techniques in economics and is used in many fields. The subtlety lies in the choice of weights, i.e. the shape of the projection function, and various approaches have been developed to find weights depending on the nature of the problem. Recent work [13] proposed to compare robustness of different aggregation techniques through sensitivity analysis, performed by Monte-Carlo simulations on synthetic data. Distribution of biases where obtained for various techniques and some showed to perform significantly better than others. Robustness assessment still depended on models used in that work.

The rest of the paper is organized as follows: Sect. 2 describes intuitively and mathematically the proposed framework; Sect. 3 then details implementation, data collection for case studies and numerical results for an artificial intra-urban case and a metropolitan real case; Sect. 4 finally discuss limitations and potentialities of the method.

2 Framework Description

2.1 Intuitive Description

We describe now the abstract framework allowing theoretically to compare robustnesses of evaluations of two different urban systems. Our framework is a generalization of an empirical method proposed in [14] besides a more general benchmarking study on indicator sense and relevance in a sustainability context. Intuitively, it relies on empirical base resulting from the following axioms:

- Urban systems can be seen from the information available, i.e. raw data describing the system. As a data-driven approach, this raw data is the basis of our framework and robustness will be determined by its structure.
- From data are computed indicators (objective functions). We assume that a choice of indicators is an intention to translate particular aspects of the system, i.e. to capture a realization of an "urban fact" (*fait urbain*) in the sense of MANGIN [15]—a sort of stylized fact in terms of processes and mechanisms, having various realizations on spatially distinct systems, depending on each precise context.
- Given many systems and associated indicators, a common space can be built to compare them. In that space, data represents more or less well real systems, depending e.g. on initial scale, precision of data, missing data. We precisely propose to capture that through the notion of point cloud discrepancy, which is a mathematical tool coming from sampling theory expressing how a dataset is distributed in the space it is embedded in [16].

Synthesizing these requirements, we propose a notion of *Robustness* of an evaluation that captures both, by combining data reliability with relative importance,

1. *Missing Data*: an evaluation based on more refined datasets will naturally be more robust.
2. *Indicator importance*: indicators with more relative influence will weight more on the total robustness.

2.2 Formal Description

Indicators Let $(S_i)_{1 \leq i \leq N}$ be a finite number of geographically disjoints territorial systems, that we assume described through raw data and intermediate indicators, yielding $S_i = (\mathbf{X}_i, \mathbf{Y}_i) \in \mathcal{X}_i \times \mathcal{Y}_i$ with $\mathcal{X}_i = \prod_k \mathcal{X}_{i,k}$ such that each subspace contain real matrices: $\mathcal{X}_{i,k} = \mathbb{R}^{n^X_{i,k} p^X_{i,k}}$ (the same holding for $\mathcal{Y}_i$). We also define an ontological index function $I_X(i,k)$ (resp. $I_Y(i,k)$) taking integer values which coincide if and only if the two variables have the same ontology in the sense of [17], i.e. they are supposed to represent the same real object. We distinguish "raw data" $\mathbf{X}_i$ from which indicators are computed via explicit deterministic functions, from "intermediate indicators" $\mathbf{Y}_i$

that are already integrated and can be e.g. outputs of elaborated models simulating some aspects of the urban system. We define the partial characteristic space of the "urban fact" by

$$(\mathcal{X}, \mathcal{Y}) \underset{def}{=} \left(\prod \tilde{\mathcal{X}}_c\right) \times \left(\prod \tilde{\mathcal{Y}}_c\right) = \left(\prod_{\mathcal{X}_{i,k} \in \mathcal{D}_\mathcal{X}} \mathbb{R}^{p^X_{i,k}}\right) \times \left(\prod_{\mathcal{Y}_{i,k} \in \mathcal{D}_\mathcal{Y}} \mathbb{R}^{p^Y_{i,k}}\right) \quad (1)$$

with $\mathcal{D}_\mathcal{X} = \{\mathcal{X}_{i,k} | I(i,k) \text{ distincts}, n^X_{i,k} \text{maximal}\}$ (the same holding for $\mathcal{Y}_i$). It is indeed the abstract space on which indicators are integrated. The indices c introduced as a definition here correspond to different indicators across all systems. This space is the minimal space common to all systems allowing a common definition for indicators on each.

Let $\mathbf{X}_{i,c}$ be the data canonically projected in the corresponding subspace, well defined for all i and all c. We make the key assumption that all indicators are computed by integration against a certain kernel, i.e. that for all c, there exists H_c space of real-valued functions on $(\tilde{\mathcal{X}}_c, \tilde{\mathcal{Y}}_c)$, such that for all $h \in H_c$:

1. h is "enough" regular (tempered distributions e.g.)
2. $q_c = \int_{(\tilde{\mathcal{X}}_c, \tilde{\mathcal{Y}}_c)} h$ is a function describing the "urban fact" (the indicator in itself)

Typical concrete example of kernels can be:

- A mean of rows of $\mathbf{X}_{i,c}$ is computed with $h(x) = x \cdot f_{i,c}(x)$ where $f_{i,c}$ is the density of the distribution of the assumed underlying variable.
- A rate of elements respecting a given condition C, $h(x) = f_{i,c}(x)\chi_{C(x)}$
- For already aggregated variables $\mathbf{Y}$, a Dirac distribution allows to express them also as a kernel integral.

Aggregation Weighting objectives in multi-attribute decision-making is indeed the crucial point of the processes, and numerous methods are available (see [18] for a review for the particular case of sustainable energy management). Let define weights for the linear aggregation. We assume the indicators normalized, i.e. $q_c \in [0, 1]$, for a more simple construction of relative weights. For i, c and $h_c \in H_c$ given, the weight $w_{i,c}$ is simply constituted by the relative importance of the indicator $w^L_{i,c} = \frac{\hat{q}_{i,c}}{\sum_c \hat{q}_{i,c}}$ where $\hat{q}_{i,c}$ is an estimator of q_c for data $\mathbf{X}_{i,c}$ (i.e. the effectively calculated value). Note that this step can be extended to any sets of weight attributions, by taking for example $\tilde{w}_{i,c} = w_{i,c} \cdot w'_{i,c}$ if $\mathbf{w}'$ are the weights attributed by the decision-maker. We focus here on the relative influence of attributes and thus choose this simple form for weights.

Robustness Estimation The scene is now set up to be able to estimate the robustness of the evaluation done through the aggregated function. Therefore, we apply an integral approximation method similar to methods introduced in [19], since the integrated form of indicators indeed brings the benefits of such powerful theoretical results. Let $\mathbf{X}_{i,c} = (\vec{X}_{i,c,l})_{1 \leq l \leq n_{i,c}}$ and $D_{i,c} = Disc_{\tilde{\mathcal{X}}_c, L^2}(\mathbf{X}_{i,c})$ the discrepancy of data points cloud[2] [20]. With $h \in H_c$, we have the upper bound on the integral approximation error

$$\left\| \int h_c - \frac{1}{n_{i,c}} \sum_l h_c(\vec{X}_{i,c,l}) \right\| \leq K \cdot |||h_c||| \cdot D_{i,c}$$

where K is a constant independent of data points and objective function. It directly yields

$$\left\| \int \sum w_{i,c} h_c - \frac{1}{n_{i,c}} \sum_l w_{i,c} h_c(\vec{X}_{i,c,l}) \right\| \leq K \sum_c |w_{i,c}| \; |||h_c||| \cdot D_{i,c}$$

Assuming the error reasonably realized ("worst case" scenario for knowledge of the theoretical value of aggregated function), we take this upper bound as an approximation of its magnitude. Furthermore, taking normalized indicators implies $|||h_c||| = 1$. We propose then to compare error bounds between two evaluations. They depend only on data distribution (equivalent to *statistical robustness*) and on indicators chosen (sort of *ontological robustness*, i.e. do the indicators have a real sense in the chosen context and do their values make sense), and are a way to combine these two type of robustnesses into a single value.

We thus define a *robustness ratio* to compare the robustness of two evaluations by

$$R_{i,i'} = \frac{\sum_c w_{i,c} \cdot D_{i,c}}{\sum_c w_{i',c} \cdot D_{i',c}} \tag{2}$$

The intuitive sense of this definition is that one compares robustness of evaluations by comparing the highest error done in each based on data structure and relative importance.

By taking then an order relation on evaluations by comparing the position of the ratio to one, it is obvious that we obtain a complete order on all possible evaluations. This ratio should theoretically allow to compare any evaluation of an urban system. To keep an ontological sense to it, it should be used to compare disjoints sub-systems with a reasonable proportion of indicators in common, or the same sub-system with varying indicators. Note that it provides a way to test the influence of indicators on an evaluation by analyzing the sensitivity if the ratio to their removal. On the contrary, finding a "minimal" number of indicators each making the ratio strongly vary should be a way to isolate essential parameters ruling the sub-system.

---

[2]The discrepancy is defined as the $L2$-norm of local discrepancy which is for normalized data points $\mathbf{X} = (x_{ij}) \in [0, 1]^d$, a function of $\mathbf{t} \in [0, 1]^d$ comparing the number of points falling in the corresponding hypercube with its volume, by $disc(\mathbf{t}) = \frac{1}{n} \sum_i \mathbb{1}_{\prod_j x_{ij} < t_j} - \prod_j t_j$. It is a measure of how the point cloud covers the space.

3 Results

Implementation Preprocessing of geographical data is made through QGIS [21] for performance reasons. Core implementation of the framework is done in R [22] for the flexibility of data management and statistical computations. Furthermore, the package `DiceDesign` [23] written for numerical experiments and sampling purposes, allows an efficient and direct computation of discrepancies. Last but not least, all source code is openly available on the `git` repository of the project[3] for reproducibility purposes [24].

3.1 Implementation on Synthetic Data

We propose in a first time to illustrate the implementation with an application to synthetic data and indicators, for intra-urban quality indicators in the city of Paris.

Data Collection We base our virtual case on real geographical data, in particular for *arrondissements* of Paris. We use open data available through the OpenStreetMap project [25] that provides accurate high definition data for many urban features. We use the street network and position of buildings within the city of Paris. Limits of *arrondissements*, used to overlay and extract features when working on single districts, are also extracted from the same source. We use centroids of buildings polygons, and segments of street network. Dataset overall consists of around 200k building features and 100k road segments.

Virtual Cases We work on each district of Paris (from the 1st to the 20th) as an evaluated urban system. We construct random synthetic data associated to spatial features, so each district has to be evaluated many time to obtain mean statistical behavior of toy indicators and robustness ratios. The indicators chosen need to be computed on residential and street network spatial data. We implement two mean kernels and a conditional mean to show different examples, linked to environmental sustainability and quality of life, that are required to be maximized. Note that these indicators have a real meaning but no particular reason to be aggregated, they are chosen here for the convenience of the toy model and the generation of synthetic data. With $a \in \{1 \dots 20\}$ the number of the district, $A(a)$ corresponding spatial extent, $b \in B$ building coordinates and $s \in S$ street segments, we take

- Complementary of the average daily distance to work with car per individual, approximated by, with $n_{cars}(b)$ number of cars in the building (randomly generated by associated of cars to a number of building proportional to motorization rate α_m 0.4 in Paris), d_w distance to work of individuals (generated from the building to a uniformly generated random point in spatial extent of the dataset), and d_{max} the diameter of Paris area, $\bar{d}_w = 1 - \frac{1}{|b \in A(a)|} \cdot \sum_{b \in A(a)} n_{cars}(b) \cdot \frac{d_w}{d_{max}}$

[3] at https://github.com/JusteRaimbault/RobustnessDiscrepancy.

- Complementary of average car flows within the streets in the district, approximated by, with $\varphi(s)$ relative flow in street segment s, generated through the minimum of 1 and a log-normal distribution adjusted to have 95 % of mass smaller than 1 what mimics the hierarchical distribution of street use (corresponding to betweenness centrality), and $l(s)$ segment length, $\bar{\varphi} = 1 - \frac{1}{|s \in A(a)|} \cdot \sum_{s \in A(a)} \varphi(s) \cdot \frac{l(s)}{\max(l(s))}$
- Relative length of pedestrian streets $\bar{p}$, computed through a randomly uniformly generated dummy variable adjusted to have a fixed global proportion of segments that are pedestrian.

As synthetic data are stochastic, we run the computation for each district $N = 50$ times, what was a reasonable compromise between statistical convergence and time required for computation. Table 1 shows results (mean and standard deviations) of indicator values and robustness ratio computation. Obtained standard deviation confirm that this number of repetitions give consistent results. Indicators obtained through a fixed ratio show small variability what may a limit of this toy approach. However, we obtain the interesting result that a majority of districts give more robust evaluations than 1st district, what was expected because of the size and content of this district: it is indeed a small one with large administrative buildings, what means less spatial elements and thus a less robust evaluation following our definition of the robustness.

3.2 *Application to a Real Case: Metropolitan Segregation*

The first example was aimed to show potentialities of the method but was purely synthetic, hence yielding no concrete conclusion nor implications for policy. We propose now to apply it to real data for the example of metropolitan segregation.

Data We work on income data available for France at an intra-urban level (basic statistical units IRIS) for the year 2011 under the form of summary statistics (deciles if the area is populated enough to ensure anonymity), provided by INSEE.[4] Data are associated with geographical extent of statistical units, allowing computation of spatial analysis indicators.

Indicators We use here three indicators of segregation integrated on a geographical area. Let assume the area divided into covering units S_i for $1 \leq i \leq N$ with centroids (x_i, y_i). Each unit has characteristics of population P_i and median income X_i. We define spatial weights used to quantify strength of geographical interactions between units i, j, with d_{ij} euclidian distance between centroids: $w_{ij} = \frac{P_i P_j}{(\sum_k P_k)^2} \cdot \frac{1}{d_{ij}}$ if $i \neq i$ and $w_{ii} = 0$. The normalized indicators are the following

[4]http://www.insee.fr.

Table 1 Numerical results of simulation for each district with $N = 50$ repetitions. Each toy indicator value is given by mean on repetitions and associated standard deviation. Robustness ratio is computed relative to first district (arbitrary choice). A ratio smaller than 1 means that integral bound is smaller for upper district, i.e. that evaluation is more robust for this district. Because of the small size of first district, we expected a majority of district to give ratio smaller than 1, what is confirmed by results, even when adding standard deviations.

Arrdt	$< \bar{d}_w > \pm\sigma(\bar{d}_w)$	$< \bar{\varphi} > \pm\sigma(\bar{\varphi})$	$< \bar{p} > \pm\sigma(\bar{p})$	$R_{i,1}$
1th	0.731655 ± 0.041099	0.917462 ± 0.026637	0.191615 ± 0.052142	1.000000 ± 0.000000
2th	0.723225 ± 0.032539	0.844350 ± 0.036085	0.209467 ± 0.058675	1.002098 ± 0.039972
3th	0.713716 ± 0.044789	0.797313 ± 0.057480	0.185541 ± 0.065089	0.999341 ± 0.048825
4th	0.712394 ± 0.042897	0.861635 ± 0.030859	0.201236 ± 0.044395	0.973045 ± 0.036993
5th	0.715557 ± 0.026328	0.894675 ± 0.020730	0.209965 ± 0.050093	0.963466 ± 0.040722
6th	0.733249 ± 0.026890	0.875613 ± 0.029169	0.206690 ± 0.054850	0.990676 ± 0.031666
7th	0.719775 ± 0.029072	0.891861 ± 0.026695	0.209265 ± 0.041337	0.966103 ± 0.037132
8th	0.713602 ± 0.034423	0.931776 ± 0.015356	0.208923 ± 0.036814	0.973975 ± 0.033809
9th	0.712441 ± 0.027587	0.910817 ± 0.015915	0.202283 ± 0.049044	0.971889 ± 0.035381
10th	0.713072 ± 0.028918	0.881710 ± 0.021668	0.210118 ± 0.040435	0.991036 ± 0.038942
11th	0.682905 ± 0.034225	0.875217 ± 0.019678	0.203195 ± 0.047049	0.949828 ± 0.035122
12th	0.646328 ± 0.039668	0.920086 ± 0.019238	0.198986 ± 0.023012	0.960192 ± 0.034854
13th	0.697512 ± 0.025461	0.890253 ± 0.022778	0.201406 ± 0.030348	0.960534 ± 0.033730
14th	0.703224 ± 0.019900	0.902898 ± 0.019830	0.205575 ± 0.038635	0.932755 ± 0.033616
15th	0.692050 ± 0.027536	0.891654 ± 0.018239	0.200860 ± 0.024085	0.929006 ± 0.031675
16th	0.654609 ± 0.028141	0.928181 ± 0.013477	0.202355 ± 0.017180	0.963143 ± 0.033232
17th	0.683020 ± 0.025644	0.890392 ± 0.023586	0.198464 ± 0.033714	0.941025 ± 0.034951
18th	0.699170 ± 0.025487	0.911382 ± 0.027290	0.188802 ± 0.036537	0.950874 ± 0.028669
19th	0.655108 ± 0.031857	0.884214 ± 0.027816	0.209234 ± 0.032466	0.962966 ± 0.034187
20th	0.637446 ± 0.032562	0.873755 ± 0.036792	0.196807 ± 0.026001	0.952410 ± 0.038702

- Spatial autocorrelation Moran index, defined as weighted normalized covariance of median income by $\rho = \frac{N}{\sum_{ij} w_{ij}} \cdot \frac{\sum_{ij} w_{ij}(X_i - \bar{X})(X_j - \bar{X})}{\sum_i (X_i - \bar{X})^2}$
- Dissimilarity index (close to Moran but integrating local dissimilarities rather than correlations), given by $d = \frac{1}{\sum_{ij} w_{ij}} \sum_{ij} w_{ij} \left| \tilde{X}_i - \tilde{X}_j \right|$ with $\tilde{X}_i = \frac{X_i - \min(X_k)}{\max(X_k) - \min(X_k)}$
- Complementary of the entropy of income distribution that is a way to capture global inequalities $\varepsilon = 1 + \frac{1}{\log(N)} \sum_i \frac{X_i}{\sum_k X_k} \cdot \log\left(\frac{X_i}{\sum_k X_k}\right)$

Numerous measures of segregation with various meanings and at different scales are available, as for example at the level of the unit by comparison of empirical wage distribution with a theoretical null model [26]. The choice here is arbitrary in order to illustrate our method with a reasonable number of dimensions.

Results We apply our method with these indicators on the Greater Paris area, constituted of four *départements* that are intermediate administrative units. The recent creation of a new metropolitan governance system [27] underlines interrogations on its consistence, and in particular on its relation to intermediate spatial inequalities. We show in Fig. 1 maps of spatial distribution of median income and corresponding local index of autocorrelation. We observe the well-known West-East opposition and district disparities inside Paris as they were formulated in various studies, such as [28] through the analysis of real estate transactions dynamics. We then apply our framework to answer a concrete question that has implications for urban policy: *how are the evaluation of segregation within different territories sensitive to missing data?* To

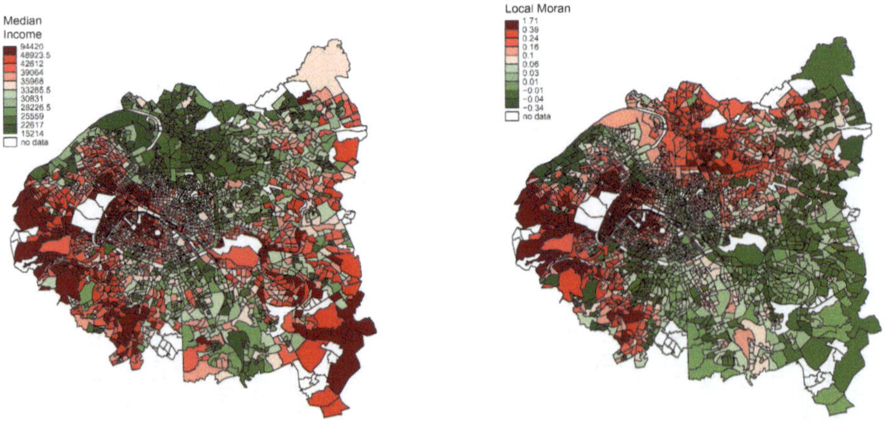

Fig. 1 Maps of metropolitan segregation. Maps show yearly median income on basic statistical units (IRIS) for the three departments constituting mainly the Great Paris metropolitan area, and the corresponding local Moran spatial autocorrelation index, defined for unit i as $\rho_i = N / \sum_j w_{ij} \cdot \frac{\sum_j w_{ij}(X_j - \bar{X})(X_i - \bar{X})}{\sum_i (X_i - \bar{X})^2}$. The most segregated areas coincide with the richest and the poorest, suggesting an increase of segregation in extreme situations

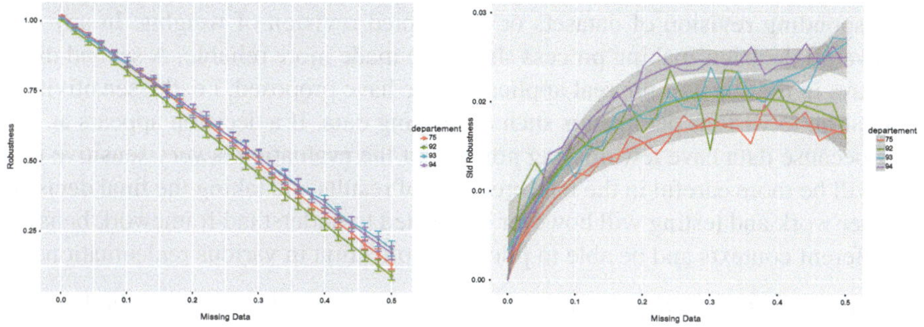

Fig. 2 Sensitivity of robustness to missing data. *Left.* For each department, Monte Carlo simulations (N = 75 repetitions) are used to determine the impact of missing data on robustness of segregation evaluation. Robustness ratios are all computed relatively to full metropolitan area with all available data. Quasi-linear behavior translates an approximative linear decrease of discrepancy as a function of data size. The similar trajectory of poorest departments (93,94) suggest the correction to linear behavior being driven be segregation patterns. *Right.* Corresponding standard deviations of robustness ratios. Different regimes (in particular 93 against others) unveil phase transitions at different levels of missing data, meaning that the evaluation in 94 is from this point of view more sensitive to missing data

do so, we proceed to Monte Carlo simulations (75 repetitions) during which a fixed proportion of data is randomly removed, and the corresponding robustness index is evaluated with renormalized indicators. Simulations are done on each *department* separately, each time relatively to the robustness of the evaluation of full Greater Paris. Results are shown in Fig. 2. All areas present a slightly better robustness than the reference, what could be explained by local homogeneity and thus more fiable segregation values. Implications for policy that can be drawn are for example direct comparisons between areas: a loss of 30 % of information on 93 area corresponds to a loss of only 25 % in 92 area. The first being a deprived area, the inequality is increased by this relative lower quality of statistical information. The study of standard deviations suggest further investigations as different response regimes to data removal seem to exist.

4 Discussion

4.1 Applicability to Real Situations

Implications for Decision-making The application of our method to concrete decision-making can be thought in different ways. First in the case of a comparative multi-attribute decision process, such as the determination of a transportation corridor, the identification of territories on which the evaluation may be flawed (i.e. has a poor relative robustness) could allow a more refined focus on these and a corresponding revision of datasets or an adapted revision of weights. In any case the overall decision-making process should be made more reliable. A second direction lays in the spirit of the real application we have proposed, i.e. the sensitivity of evaluation to various parameters such as missing data. If a decision appears as reliable because data have few missing points, but the evaluation is very sensitive to it, one will be more careful in the interpretation of results and taking the final decision. Further work and testing will however be needed to understand framework behavior in different contexts and be able to pilot its application in various real situations.

Integration Within Existing Frameworks The applicability of the method on real cases will directly depend on its potential integration within existing framework. Beyond technical difficulties that will surely appear when trying to couple or integrate implementations, more theoretical obstacles could occur, such as fuzzy formulations of functions or data types, consistency issues in databases, etc. Such multicriteria framework are numerous. Further interesting work would be to attempt integration into an open one, such as e.g. the one described in [29] which calculates various indices of urban segregation, as we have already illustrated the application on metropolitan segregation indexes.

Availability of Raw Data In general, sensitive data such as transportation questionnaires, or very fine granularity census data are not openly available but provided already aggregated at a certain level (for instance French Insee Data are publicly available at basic statistical unit level or larger areas depending on variables and minimal population constraints, more precise data is under restricted access). It means that applying the framework may imply complicated data research procedure, its advantage to be flexible being thus reduced through additional constraints.

4.2 Validity of Theoretical Assumptions

A possible limitation of our approach is the validity of the assumption formulating indicators as spatial integrals. Indeed, many socio-economic indicators are not necessarily depending explicitly on space, and trying to associate them with spatial coordinates may become a slippery slope (e.g. associate individual economic variables with individual residential coordinates will have a sense only if the use of the variable has a relation with space, otherwise it is a non-legitimate artifact). Even indicators which have a spatial value may derive from non-spatial variables, as [30] points out concerning accessibility, when opposing integrated accessibility measures with individual-based non necessarily spatial-based (e.g. individual decisions) measures. Constraining a theoretical representation of a system to fit a framework by changing some of its ontological properties (always in the sense of real meaning of objects) can be understood as a violation of a fundamental rule of modeling and simulation in social science given in [31], that is that there can be an universal "language" for modeling and some can not express some systems, having for consequence misleading conclusion due to ontology breaking in the case of an over-constrained formulation.

4.3 Framework Generality

We argue that the fundamental advantage of the proposed framework is its generality and flexibility, since robustness of the evaluations are obtained only through data structure if ones relaxes constraints on the value of weight. Further work should go towards a more general formulation, suppressing for example the linear aggregation assumption. Non-linear aggregation functions would require however to present particular properties regarding integral inequalities. For example, similar results could search in the direction of integral inequalities for Lipschitzian functions such as the one-dimensional results of [32].

5 Conclusion

We have proposed a model-independent framework to compare the robustness of multi-attribute evaluations between different urban systems. Based on data discrepancy, it provide a general definition of relative robustness without any assumption on model for the system, but with limiting assumptions that are the need of linear aggregation and of indicators being expressed through spatial kernel integrals. We propose a toy implementation based on real data for the city of Paris, numerical results confirming general expected behavior, and an implementation on real data for income segregation on Greater Paris metropolitan areas, giving possible insights into concrete policy questions. Further work should be oriented towards sensitivity analysis of the method, application to other real cases and theoretical assumptions relaxation, i.e. the relaxation of linear aggregation and spatial integration.

Acknowledgements The author would like to thank Julien Keutchayan (Ecole Polytechnique de Montréal) for suggesting the original idea of using discrepancy, and anonymous reviewers for the useful comments and insights.

References

1. Marler, R.T., Arora, J.S.: Survey of multi-objective optimization methods for engineering. Struct. Multidiscip. Optim. **26**(6), 369–395 (2004)
2. Newman, M.E.J.: Complex systems: a survey (2011). arXiv:1112.1440
3. Haken, H., Portugali, J.: The face of the city is its information. J. Env. Psychol. **23**(4), 385–408 (2003)
4. Picon, A.: Smart cities: théorie et critique d'un idéal auto-réalisateur. B2 (2013)
5. Souami, T.: Ecoquartiers: secrets de fabrication. Scrineo (2012)
6. Bavoux, J.-J., Beaucire, F., Chapelon, L., Zembri, P.: Géographie des transports. Paris (2005)
7. Carver, J.S.: Integrating multi-criteria evaluation with geographical information systems. Int. J. Geogr. Inf. Syst. **5**(3), 321–339 (1991)

8. Jégou, A., Augiseau, V., Guyot, C., Judéaux, C., Monaco, F.-X., Pech, P. et al.: L'évaluation par indicateurs: un outil nécessaire d'aménagement urbain durable?. réflexions à partir de la démarche parisienne pour le géographe et l'aménageur. Cybergeo: European Journal of Geography (2012)
9. Launer, R.L., Wilkinson, G.N.: Robustness in statistics. Academic Press (2014)
10. Deb, K., Gupta, H.: Introducing robustness in multi-objective optimization. Evol. Comput. **14**(4), 463–494 (2006)
11. Barrico, C., Antunes, C.H.: Robustness analysis in multi-objective optimization using a degree of robustness concept. In: Evolutionary Computation, 2006. CEC 2006. IEEE Congress on, pp. 1887–1892 (2006)
12. Brunsdon, C., Fotheringham, S., Charlton, M.: Geographically weighted regression. J. R. Stat. Soc. Ser. D (The Statistician) **47**(3), 431–443 (1998)
13. Dobbie, M.J., Dail, D.: Robustness and sensitivity of weighting and aggregation in constructing composite indices. Ecol. Indic. **29**, 270–277 (2013)
14. Ali, A., Carneiro, I., Dussarps, L., Guédel, F., Lamy, E., Raimbault, J., Viger, L., Cohen, V., Aw, T., Sadeghian, S.: Les eco-quartiers lus par la mobilité : vers une évaluation intégrée. Technical report, Ecole des Ponts ParisTech (2014). June
15. Mangin, D., Panerai, P.: Projet urbain. Parenthèses (1999)
16. Dick, J., Pillichshammer, F.: Digital Nets and Sequences: Discrepancy Theory and Quasi–Monte Carlo Integration. Cambridge University Press (2010)
17. Livet, P., Muller, J.-P., Phan, D., Sanders, L.: Ontology, a mediator for agent-based modeling in social science. J. Artif. Soc. Soc. Simul. **13**(1), 3 (2010)
18. Wang, J.-J., Jing, Y.-Y., Zhang, C.-F., Zhao, J.-H.: Review on multi-criteria decision analysis aid in sustainable energy decision-making. Renew. Sustain. Energy Rev. **13**(9), 2263–2278 (2009)
19. Varet, S.: Développement de méthodes statistiques pour la prédiction d'un gabarit de signature infrarouge. Ph.D. thesis, Université Paul Sabatier-Toulouse III (2010)
20. Niederreiter, H.: Discrepancy and convex programming. Annali di matematica pura ed applicata **93**(1), 89–97 (1972)
21. DT QGis. Quantum gis geographic information system. Open Source Geospatial Foundation Project (2011)
22. R Core Team: R language definition (2000)
23. Jessica, F., Delphine, D., Olivier, R., Astrid, J.: Dicedesign-package. Designs of Computer Experiments, pp. 2 (2009)
24. Ram, K.: Git can facilitate greater reproducibility and increased transparency in science. Source Code Biol. Med. **8**(1), 7 (2013)
25. Bennett, J.: OpenStreetMap. Packt Publishing Ltd. (2010)
26. Louf, R., Barthelemy, M.: Patterns of residential segregation (2015). arXiv:1511.04268
27. Gilli, F., Offner, J.-M.: Paris, métropole hors les murs: aménager et gouverner un Grand Paris. Sciences Po, les presses (2009)
28. Guérois, M., Le Goix, R.: La dynamique spatio-temporelle des prix immobiliers à différentes échelles: le cas des appartements anciens à paris (1990–2003). Cybergeo: European Journal of Geography (2009)
29. Tivadar, M., Schaeffer, Y., Torre, A., Bray, F.: Oasis–un outil d'analyse de la ségrégation et des inégalités spatiales. Cybergeo: European Journal of Geography (2014)
30. Kwan, M.-P.: Space-time and integral measures of individual accessibility: a comparative analysis using a point-based framework. Geogr. Anal. **30**(3), 191–216 (1998)
31. Banos, A.: Pour des pratiques de modélisation et de simulation libérées en géographie et shs. *Thèse d'Habilitation à Diriger des Recherches, UMR CNRS 8504 Géographie-Cités, ISCPIF*, Décembre (2013)
32. Dragomir, S.S.: The ostrowski's integral inequality for lipschitzian mappings and applications. Comput. Math. Appl. **38**(11), 33–37 (1999)

Complexity Management for Engineered Systems Using System Value Definition

Kaushik Sinha, Narek R. Shougarian and Olivier L. de Weck

Abstract Quantitative and objective management of complexity is essential for effective design of engineered complex systems. In this paper, we develop a quantitative framework for complexity management. This includes a measure of system value that explicitly considers system complexity. The system design goal is to maximize the system value. Using a simple, representative mathematical model linking performance to system complexity, we show analytically that there exists a regime where we have an optimal level of complexity that leads to maximization of system value. Existence of this regime is dependent on two rate parameters that link the complexity-performance-development cost triad for engineered systems. Outside of this regime one has to always aim for reducing system complexity in order to maximize system value. The framework is subsequently applied to a case study involving a set of aircraft engine architectures.

Keywords Complexity management · System performance · System development cost · System value · Performance gain · Complexity penalty · Complexity budget · System value maximization · Aircraft engine architectures

Nomenclature

N	Number of Components
M	Number of Interfaces

K. Sinha (✉)
Institute for Data, Systems and Society, Massachusetts Institute of Technology, Cambridge, MA 02139, USA
e-mail: sinhak@mit.edu

N.R. Shougarian
Department of Aeronautics and Astronautics, Massachusetts Institute of Technology, Cambridge, MA 02139, USA
e-mail: nshoug@mit.edu

O.L. de Weck
Department of Aeronautics and Astronautics and Institute for Data, Systems and Society, Massachusetts Institute of Technology, Cambridge, MA 02139, USA
e-mail: deweck@mit.edu

G. Fanmuy et al. (eds.), *Complex Systems Design & Management*,
DOI 10.1007/978-3-319-49103-5_12

α_i	Component Complexity
β_{ij}	Interface Complexity
A	Binary Adjacency Matrix
$E(A)$	System Graph Energy
C	Structural Complexity
$\sigma_i(.)$	Singular Value
P	System Performance
NRE	Non-Recurring Engineering Cost
(n,k)	Complexity-Performance Tradespace Parameters
(a,m)	Complexity-NRE Tradespace Parameters
V	System Value

1 Introduction

Most modern era software-enabled, electro-mechanical systems are becoming more and more complex as we demand more performance and better lifecycle properties (e.g. robustness) [1–3]. As a consequence system development projects are becoming increasingly challenging and are falling behind in terms of schedule and cost performance. There is consensus that this is due to our poor understanding of how to measure and manage complexity [4, 5]. Although complexity has received widespread attention, yet there is still significant work to be done in bridging quantification of complexity and its implication for system management. Complexity Management refers to maintaining balance between system's performance, its complexity and the system development cost. From a system design perspective, complexity reduction is not an end goal in itself. The end goal is to provide target system performance level while minimizing complexity and development cost. Alternatively, we can term the end goal as maximizing system value, defined as achieved performance per unit development cost/effort. In order to achieve this end goal, how should we actively manage the underlying system complexity? This is the primary question this paper aims to address. In this paper, we formulate a system value evaluation model that encompasses the system performance, complexity and development cost triad. This methodology enables active complexity management and maximizes the system value. This framework brings forth the notion of *complexity budgeting*, similar in spirit to the mass budget or power budget used, for example, in aerospace system development projects. We demonstrate that the system value versus complexity trade space can be divided into two regimes with very different characteristics based on a ratio of two parameters that governs the performance-complexity-development cost triad. We subsequently apply the framework to a set of aircraft engine architectures [6, 7].

2 Complexity Management Framework

This section discusses complexity quantification, system performance measurement and complexity-enabled system value definition which constitute the complexity management framework and subsequently optimize for system value.

2.1 Complexity Quantification

The structural complexity of engineering systems depends on the number and characteristics of different elements, their connectivity structure and is a measurable system characteristic. The structural complexity metric used in this paper include contributions coming from the internal complexities of the components of the system; the complexities associated to the individual connections among the components and a quantity that encapsulates the complexity due to inherent arrangement of connections (i.e., structure) amongst the components. Note that the system behavior is not explicitly considered in estimation of structural complexity. The functional form that we adopted [1, 8] for estimating the structural complexity of an engineered complex system is,

$$\text{Structural Complexity, } C = C_1 + C_2 C_3$$

Here the first term C_1 represents the sum of complexities of individual components. The individual component complexities can vary across the system (e.g., a low-pressure turbine is much more complex than the exhaust nozzle in a jet engine) and are designated by α_i:

$$C_1 = \sum_{i=1}^{N} \alpha_i$$

Notice that this term, C_1 represents component complexity alone and does not involve architectural information. This is indicative of component or technology development efforts and therefore uses information local to the component being developed (or acquired from a catalogue of components). Estimation of component complexity can be pursued in multiple ways depending on the availability of the amount of data. We can adopt more qualitative scale-based measures like Technology Readiness Levels (TRL), to a more data-centric method that depends on the availability of historical data on similar components or a hybrid methodology that includes expert-opinion coupled with historical data [8]. The important aspect is to use consistent estimation methods while comparing different system architectures. The second term C_2 represents the sum of complexities of each pair-wise interaction β_{ij}:

$$C_2 = \sum_{i=1}^{N} \sum_{j=1}^{N} \beta_{ij} A_{ij}$$

where, $A \in M_{N \times N}$ represents the binary adjacency matrix representing the connectivity structure of the system:

$$A_{ij} = \begin{cases} 1 & \forall[(i,j)|(i \neq j) \text{ and } (i,j) \in \Lambda] \\ 0 & \text{otherwise} \end{cases}$$

where Λ represents the set of connected nodes and N being the number of components in the system. Each non-zero entry in the binary adjacency matrix A_{ij} represents a connection from the component j to the component i. The diagonal elements of A are zero by construction. Notice that the interface complexity term C_2 involves quantification/categorization of each pair-wise interface and requires the partial knowledge about the system architecture while estimating β_{ij}. Estimation of interface complexity also depends on the amount of available data and the range of possible estimation approaches can be found in [8]. This component of structural complexity is indicative of interface development and management efforts.

The third term C_3 represents the topological arrangement of the interfaces and is called the *topological complexity* metric, which is given as:

$$C_3 = \frac{E(A)}{N}, \quad \text{where } E(A) = \sum_{i=1}^{N} \sigma_i(A)$$

where $\sigma_i(.)$ represents the *i*th singular value of the binary adjacency matrix A. Notice that the definition of graph energy $E(A)$ involves singular values of the adjacency matrix which are real, non-negative (i.e., $\sigma_i(.) \geq 0, \forall i$) for any general binary adjacency matrix. This quantity is well-defined, non-negative number for both symmetric and asymmetric matrices and therefore, can handle both undirected and directed interfaces.

Notice that C_3 requires knowledge of the overall system architecture (i.e., knowledge of the system level binary adjacency matrix A) and in this sense, represents a global effect whose impact could be realized at the time of system integration. The topological complexity C_3 helps distinguish structural complexity of very different connectivity structures with the same number of components and interactions (see Fig. 1). The second term C_2C_3 in Eq. (1) is an overall indicator of system integration (or system realization) effort.

Fig. 1 Two architectures having the same number of nodes and connections but are differentiated based on their internal structure with $E(A_1) = 4.9$ and $E(A_2) = 6.83$

The structural complexity metric introduced in detailed form is shown below:

$$C = \sum_{i=1}^{N} \alpha_i + \left(\sum_{i=1}^{N} \sum_{j=1}^{N} \beta_{ij} A_{ij} \right) \frac{E(A)}{N}$$

Implication of different terms of the structural complexity metric is described in Fig. 2.

The structural complexity metric introduced above has been checked for conceptual validity against the set of required properties prescribed by Weyuker [9] and shown to be fully compliant with Weyuker's criteria [1, 8]. For brevity, we do not discuss additional details about model-based estimation of component and interface complexities and important properties of the structural complexity in the context of system architecting. Interested readers can refer to more detailed exposition of this metric elsewhere [8].

2.2 *System Performance*

The primary "Technical Performance Measure" (TPM) we use in this paper (for aircraft engines) is "Thrust Specific Fuel Consumption" (TSFC) which is defined as the ratio between engine fuel consumption and its thrust. We approximate engine behavior using a number of steady state operating points at different altitudes and Mach numbers and the TSFC was calculated at multiple points of a representative commercial aircraft flight profile. Propulsion system performance was computed

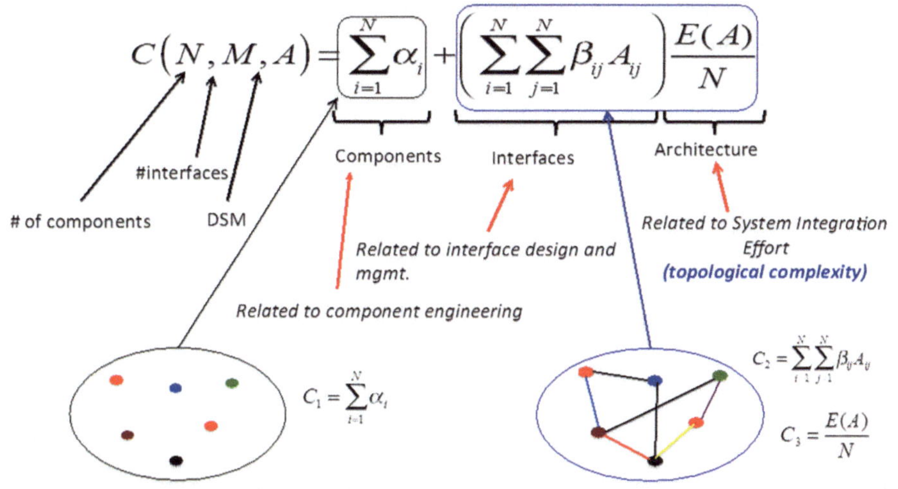

Fig. 2 Constituents of different parts of the overall structural complexity metric and their implications in the context of system development

using the well-known Gasturb11 software. Gasturb11 leverages 1D thermodynamic relations and a library of compressor, fan and turbine performance maps to compute steady state design point and off-design engine performance. Gasturb11 also allows transient behavior of engines to be computed. Details of the performance measure are discussed later in this paper.

2.3 Complexity-Enabled System Value

We adopt a simplified S-curve model for representing the system performance and complexity relationship, based on observations made in the literature [7, 10]. Also note that this model represents an envelope of "well engineered" systems architectures (i.e., they do not contain additional complexity that does not bring in performance benefits). We can treat this envelope as an approximate Pareto-Optimal curve in performance-complexity trade space. This model shows that for well-engineered systems, we achieve performance gains with increasing complexity, but the benefit tapers off beyond a certain level of complexity (i.e., diminishing *performance returns* on *complexity investments*).

The relationship between system performance and complexity can be modeled using S-curve with parameters (n, k) as shown below [1, 7],

$$P = P_{\max}\left(\frac{kC^n}{1+kC^n}\right). \tag{1}$$

The system performance level is assumed to saturate at P_{max} (see Fig. 3).

As can be seen by analyzing Eq. 1, higher n indicates higher rate of performance gain and saturation at a lower complexity level, while lower value of k shifts the curve towards the right (see Fig. 4).

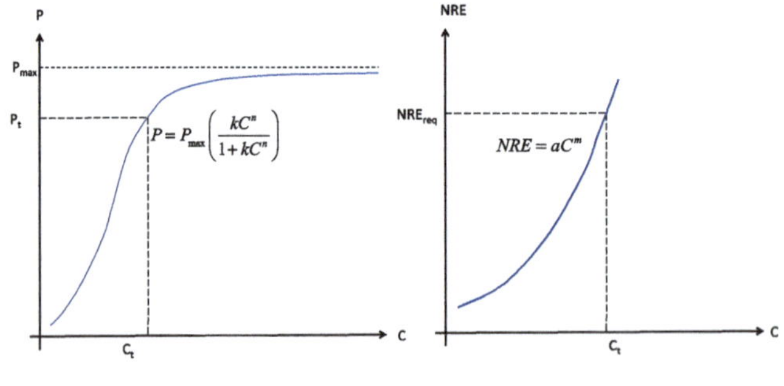

Fig. 3 The relationship between complexity and (i) performance; and (ii) NRE cost/effort

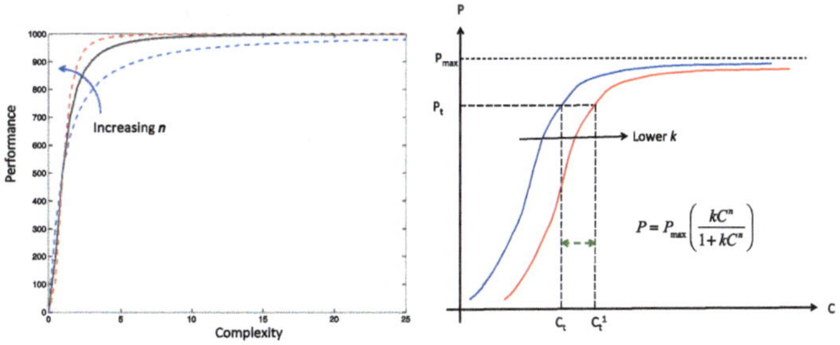

Fig. 4 (i) Increasing n means higher rate of performance gain and performance saturation at a lower complexity level; (ii) lower k shift the performance-complexity curve to the *right*, indicating higher complexity level to attain the same performance level

The Non-Recurring-Engineering (NRE) cost/effort can be well estimated by a monotonically increasing, nonlinear curve [1],

$$NRE = aC^m \tag{2}$$

where (a, m) are the parameters (see Fig. 3).

We define a value function, ***V*** that expresses the performance gain per unit NRE expenditure. This is much like a price for enhanced performance and this price is being paid to counter increased complexity. Hence, we can interpret the value function as the complexity price for performance gain. Once performance gain saturates, any increase in complexity is counter productive as we have to pay a complexity penalty in terms of increased NRE, without extracting any/nominal performance benefits. This leads to erosion in system value.

The effect of parameter m on the NRE in Eq. 1 is shown in Fig. 5. We can interpret the exponent m as the rate of NRE penalty for increased complexity. A smaller value of m indicates better complexity handling capability.

Fig. 5 Lower m value indicates smaller NRE for the same complexity level, indicating a lower rate of NRE penalty for increased complexity

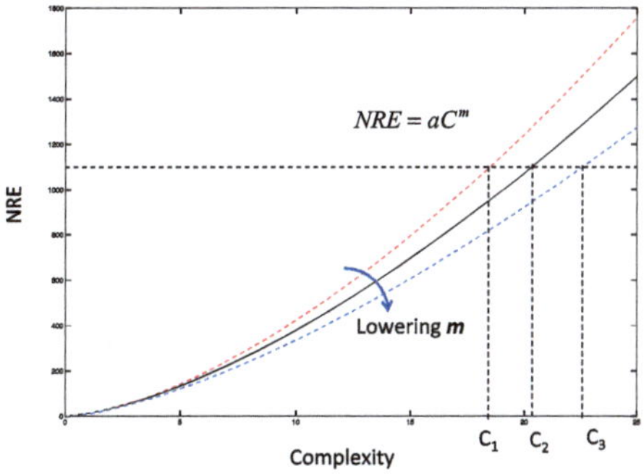

Now, using the above functional forms, the value function $\boldsymbol{V}$ can be written as,

$$V = \frac{P}{NRE} = \underbrace{P_{\max}\left(\frac{k}{a}\right)}_{S}\left[\frac{C^{(n-m)}}{1+kC^{n}}\right] \\ = S\left[\frac{C^{(n-m)}}{1+kC^{n}}\right] \tag{3}$$

where, $S = P_{\max}\left(\frac{k}{a}\right)$.

2.4 Optimizing System Value

We can observe that, for $n \leq m$, the value function decreases monotonically for increasing complexity. If we compute its first derivative with respect to complexity, we have

$$\frac{dV}{dC} = -\frac{S}{(1+kC^{n})^{2}C^{(m-n+1)}}\left[(m-n)+kmC^{n}\right] \tag{4}$$

It is clear that $\frac{dV}{dC} < 0$ for $n \leq m$. The rate of decreasing value is higher for $n < m$ and can be seen in Fig. 4. This means that, if this condition is satisfied, the value will always decrease and it does not pay (from a system value standpoint) to increase the complexity level of the system.

The *interesting* regime is when $n > m$, indicating that the initial rate for performance gain outweighs the rate of NRE penalty for increased complexity. Writing the optimality condition below, we compute the complexity level, C_* for value maximization,

$$\frac{dV}{dC} = 0 \\ \Rightarrow \frac{SC_*^{(n-m-1)}}{\left(1+kC_*^{n}\right)^{2}}\left[(n-m)+kmC_*^{n}\right] = 0 \\ C_*^{n} = \frac{\left(\frac{n}{m}\right)-1}{k} \tag{5}$$

To check if this is a maximal solution, we compute the second derivative. After algebraic simplifications, we arrive at the following expression for the second derivative at $C = C_*$,

$$\left.\frac{d^{2}V}{dC^{2}}\right|_{C_*} = \frac{-Sm^{2}(n-m)}{n}C_*^{n-m-2}$$

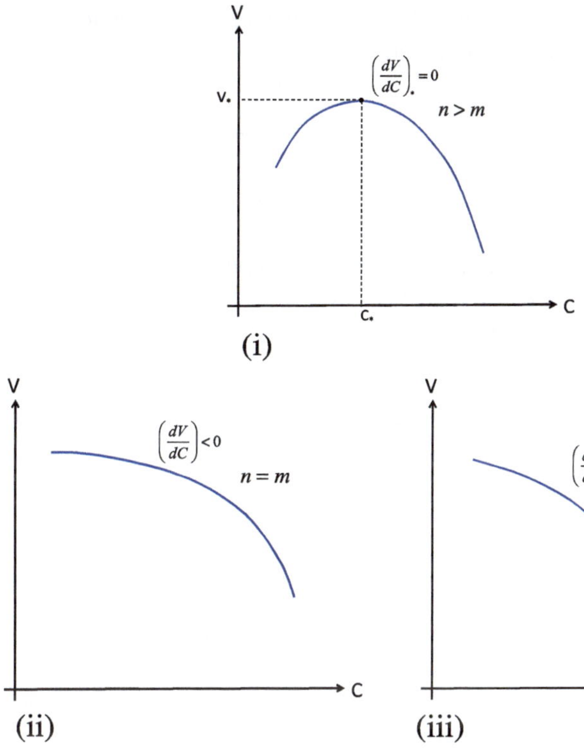

Fig. 6 Plots showing the value-complexity curves for (i) $n > m$; (ii) $n = m$; and (iii) $n < m$. For $n \leq m$, the value function is monotonically decreasing with complexity

For $n > m$, we observe that $\left.\frac{d^2V}{dC^2}\right|_{C_*} < 0$, guaranteeing that the value function V is indeed maximized at $C = C_*$. This result is graphically demonstrated in Fig. 6.

This means that, in contrast to the previous case, there is value enhancement in increasing the complexity up to a point. This is characterized by the regime $n > m$.

The corresponding performance level P_* is given by,

$$\begin{aligned} P_* &= P_{\max}\left(\frac{kC_*^n}{1 + kC_*^n}\right) \\ &= P_{\max}\frac{\left(\frac{n}{m} - 1\right)}{\left(\frac{n}{m}\right)} \\ &= P_{\max}\left(1 - \frac{m}{n}\right) \end{aligned} \tag{6}$$

Equation 6 indicates that P_* reaches $P_{\max}$ as n/m ratio increases. Large n/m ratio indicates sharper performance gain with better complexity handling/management capability. Also a smaller k combined with larger n/m ratio leads to higher complexity level, C_* at maximal value function.

The corresponding NRE value at this complexity level is given by,

$$NRE_* = aC_*^m = a\left[\frac{\left(\frac{n}{m}\right) - 1}{k}\right]^{\frac{m}{n}} \tag{7}$$

This expression is more complicated as m/n ratio appears on the exponent, but a larger n/m ratio usually leads to a smaller NRE.

The maximized system value, $V_{\max}$ is given by,

$$\begin{aligned} V_{\max} &= \frac{SC_*^{n-m}}{1+kC_*^n} \\ &= \left(\frac{m}{n}\right)SC_*^{n-m} \\ &= \left(\frac{m}{n}\right)S\left[\frac{\left(\frac{n}{m}\right) - 1}{k}\right]^{\left(1-\frac{m}{n}\right)} \end{aligned} \tag{8}$$

where, $S = P_{\max}\left(\frac{k}{a}\right)$. A large n/m ratio and low k value usually helps attain a larger maximum value function, $V_{\max}$.

From the above analysis, we can define two distinct regimes for complexity management based on the n/m ratio:

Regime I: This regime is defined by the condition $n > m$, indicating that performance gain parameter is larger than the complexity penalty parameter. This leads to existence of an optimal complexity level C_* as defined in Eq. (5) and this level of complexity is found to maximize system value as defined by Eq. (8). The system value-complexity trade space (see Fig. 6(i) above) is concave in this case with global maxima.

Regime II: This regime is complementary to regime I and is characterized by the condition $n \leq m$, indicating complexity penalty dominates over performance gain. In this case the system value-complexity trade space is monotonically decreasing (e.g., indicated by $\frac{dV}{dC} < 0$) and therefore, reducing complexity is the only way to improve the system value, while keeping other factors constant.

2.5 Complexity Budgeting

Much like the mass or power budgets used in traditional engineering system development, we can think about a notion of *complexity budget*. *Complexity budget* refers to a level of complexity that is most beneficial from a value perspective where we gain performance while keeping NRE cost/effort within prescribed/manageable limits. From a programmatic perspective, one question that remains is how to fixing

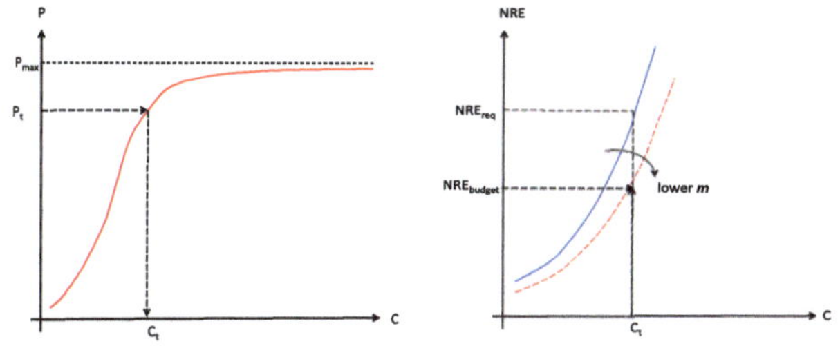

Fig. 7 Trace the trade-space for a given system performance target, P_t: (i) find the requisite complexity, C_t and (ii) find the NRE_{req} to achieve the specified level of performance for a given NRE-Complexity curve. If $\mathrm{NRE_{reg}} > \mathrm{NRE_{budget}}$, then improved complexity management is mandated to lower the m value and influence a shift in NRE-complexity profile

the desired level of complexity and how should we go about defining a *complexity budget*, just as in the case of mass budget or the power budget. For any required level of system performance, $\mathrm{P_t}$ to be achieved given the Performance-Complexity and NRE-Complexity curves, we can trace $\mathrm{NRE_{req}}$ as the expected NRE cost/effort as shown in Fig. 7.

If the budgeted NRE, $\mathrm{NRE_{budget}}$ is smaller than the estimated $\mathrm{NRE_{req}}$, then we have to look for ways to reduce the value of exponent m and shift the NRE-Complexity curve. This calls for improved complexity management strategies and this might prove to be unachievable under the circumstances. In such case, one needs to explore other options including compromising on the system performance targets or to look for different system architectures for which a higher rate of performance gain (i.e., exponent n in Eq. 1) is achievable. More detailed organizational capability based decisions on complexity management and distribution are discussed in [1].

3 Case Study: Aircraft Engine Architectures

Until now, we have developed the analytical model for complexity management framework. We adopted a simplified S-curve type model for relating system performance to its underlying complexity based on observation in existing literature [7, 10]. Such relationship has been empirically observed while studying RLC circuits [7] (see Fig. 8).

We can use a similar mathematical model (inspired from results in Fig. 8) to study engineered systems. In order to investigate this aspect empirically, we apply the proposed methodology to a set of jet engine architectures [6].

A jet engine is an air-breathing reaction engine that discharges a fast moving jet, generating thrust by jet propulsion. In general, most jet engines are internal combustion engines [6]. Jet engine architectures considered here include turbojets, turbofans and geared turbofans. Turbojets consist of an air inlet, an air compressor,

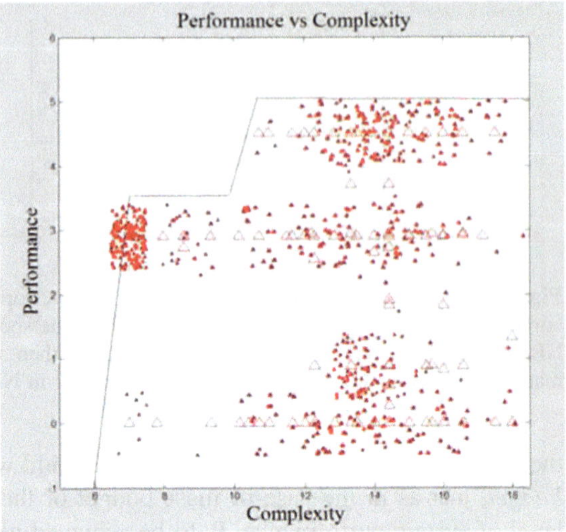

Fig. 8 Performance versus complexity curve for simple RLC circuit [11]

a combustion chamber, a gas turbine that drives the air compressor and a nozzle. The air is compressed into the chamber, heated and expanded by the fuel combustion and then allowed to expand out through the turbine into the nozzle where it is accelerated to high speed to provide propulsion. Modern subsonic jet aircraft use high-bypass turbofan engines that offer high speed with fuel efficiency (see Fig. 9). The underlying principle of the geared turbofan engine (GTF), shown in Fig. 9e, is to decouple fan and turbine speed, enabling smaller turbine designs while enabling a high bypass ratios and low fan pressure ratios. This improves propulsive efficiency, decreases noise and weight at the same time.

For the purposes of this study, we analyze a set of 5 engine architectures (see Table 2) using simplified 1D models that are simulated using GasTurb 11™ [11]. In all cases, identical flight profile as described in Table 1 with 11 segments [12], is used. We use a weighted version of thrust-specific fuel consumption (TSFC) [6] as defined below:

$$TSFC_{wt} = \sum_{i=1}^{11} \phi_{E\{i\}} TSFC_{E\{i\}}$$

We put more emphasis on cruise states since those are the states where an aircraft engine spends most of their time in flight. All 5 architectures are optimized for steady-state conditions at cruise [11].

The performance metric P is defined as:

$$P = \frac{1000}{TSFC_{wt}}$$

A higher value of P indicates superior performance.

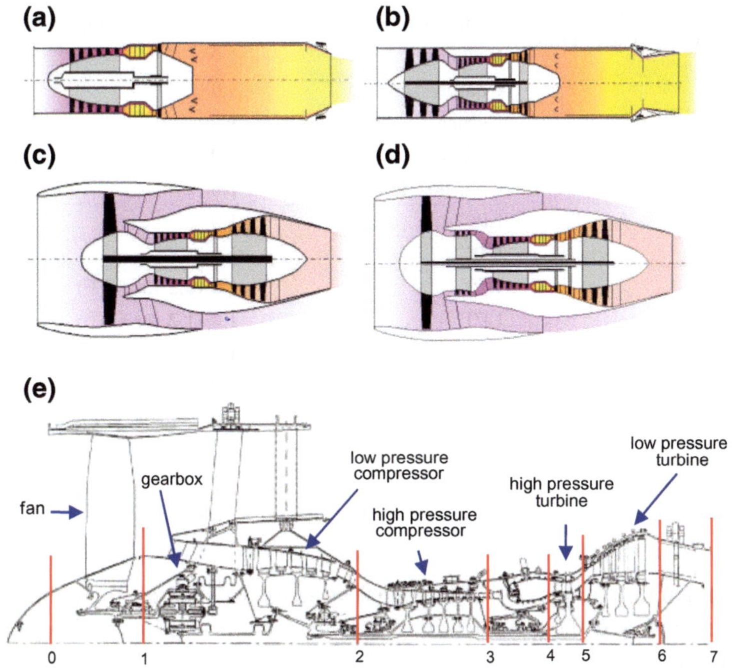

Fig. 9 **a** Turbojet; **b** 2 Spool Turbojet; **c** 2 Spool Turbofan; **d** 3 Spool Turbofan; **e** Geared Turbofan [1, 13]

Table 1 Flight profile used in simulation of all five aircraft engine architectures

Segment #	Regime	Altitude (ft)	Duration (s)	
1	Takeoff stationary	0	15	0.0011
2	Takeoff rotation	0	15	0.0011
3	Climb from 0 ft to 10000.250 knots IAS	4860	373	0.0272
4	Climb from 10000 to 29880 ft. 300 knots IAS	19620	759	0.0554
5	Climb from crossover 29880 to 36000 ft. M = 0.78	32760	248	0.0181
6	Cruise segment 1 (36000 ft)	36000	5400	0.3943
7	Cruise segment 2 (38000 ft)	38000	5400	0.3943
8	Descent from 38000 to 29640 ft. M = 0.78	33820	320	0.0234
9	Descent from 29640 to 10000 ft. 300 knots lAS	19950	743	0.0542
10	Descent from 10000 to 0 ft. 250 knots lAS	5130	393	0.0287
11	Landing	0	30	0.0022

Table 2 Component complexity values used in this example

Component name	Component complexity (α)
Intake	1.0
Compressor	1.0
Burner	1.0
Turbine	1.0
Core nozzle	1.0
LPC	1.0
HPC	1.0
HPT	1.0 (1.2 for GTF)
LPT	1.0 (1.2 for GTF)
Fan	1.0
Bypass nozzle	1.0
IPC	1.0
IPT	1.0
Booster	1.2
Gearbox	1.6

The structural complexity of technical systems depends on the quantity of different elements and their connectivity structure and is a measurable system characteristic. We use the structural complexity estimation method described earlier:

$$C = \sum_{i=1}^{n} \alpha_i + \left[\sum_{i=1}^{n} \sum_{j=1}^{n} \beta_{ij} A_{ij} \right] \frac{E(A)}{n} \tag{9}$$

In this simplified example, the component complexities used are listed in Table 2. For simplicity, we assume uniform interface complexity $\beta_{i,j} = 0.5 \ \forall (i,j)$.

For the current set of 5 architectures, we use a coarse, top-level system dependency structure [13] without the supporting systems like thermal management and active control systems. Such components will add complexity that is generally monotonic going from single spool turbojets to turbofans [6, 13] with the geared turbofan adding much higher level complexity in terms of its thermal management and active control systems [13]. Please note that this simplification, as implemented herein, does not alter the relative ranking of engine architectures in terms of their respective underlying system complexity values (Table 3).

From the Fig. 10, we observe that the performance-complexity profile for the set of jet engine architectures can be well approximated by S-curve model that we adopted earlier. This enables us to use the rest of the active complexity management framework and with system value maximization. We are currently developing a rule-based system synthesis framework that automatically generates feasible engine architectures, given a library of components that can be used to compose the system [7]. This approach enables a complete, higher fidelity characterization of system performance-complexity trade space [7].

Table 3 Architecture specific complexity and performance values

Architecture	Complexity	Performance (1000/TSFC)
1 spool turbojet	10.4	31.1
2 spool turbojet	16.02	32.7
2 spool turbofan	16.4	43.9
3 spool turbofan	23.04	47.8
Geared turbofan (GTF)	24.85	56.5

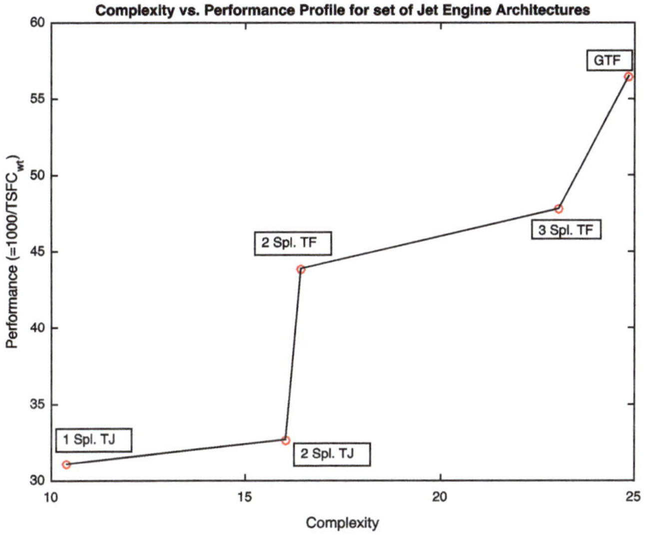

Fig. 10 Performance versus complexity profile for five aircraft engine architectures

4 Conclusions and Future Work

In this paper, we described a complexity quantification methodology, a system performance quantification process and formulated a methodology for estimating a complexity budget, similar to the notion of mass or power budgets in conventional systems engineering. Since increased complexity often enables improved system performance, we formulated a value function as amount of performance gain per unit NRE (Non-Recurring Expenditure). The system value maximization problem is characterized by two primary parameters: (i) performance gain parameter; and (ii) complexity penalty parameter. The ratio n/m is shown to define two distinct regimes from active complexity management perspective. One regime has an optimal level of complexity C_* that maximizes system value, while the other regime dictates active complexity reduction/containment to enhance/maintain the system value. We found that there exists a maximum system value (i.e., regime I) if the rate of performance gain (with increasing complexity) is greater than the rate of NRE cost/effort penalty due to the increased complexity. In all other cases, the system value is found to decrease monotonically with increasing complexity (i.e., regime II). As seen from the initial results of the engine architectures case study, the performance-complexity profile can be adequately modeled using the proposed simplified S-curve and the active complexity management framework can be applied for such a challenging engineered system that involves multiple engineering domains. We are currently evolving a rule-based architecture enumeration engine for automatic synthesis of feasible top-level jet engine architectures, given a library of components [7] to enable efficient exploration and synthesis of design architecture space. This approach enables characterization of system performance-complexity trade space at much higher granularity and will be elaborated in a future publication. In future, we plan to extend and apply this framework to different types of engineered systems spanning a diverse set of application domains.

References

1. Sinha, K.: Structural complexity and its implications for design of cyber-physical systems. Ph. D. Thesis, MIT (2014)
2. Sturtevant, D.: System design and the cost of architectural complexity. Ph.D. Thesis, MIT (2013)
3. Sinha, K., de Weck, O.L.: Structural complexity quantification for engineered complex systems and implications on system architecture and design. In: Proceedings of the International Design Engineering Technical Conference IDETC2013, August Portland, Oregon, USA, pp. 7–4 (2013)
4. de Weck, O., Roos, D., Magee, C.: Engineering Systems: Meeting Human Needs in a Complex Technological World. MIT Press (2011)
5. Crawley, E.: "Lecture notes for ESD.34—System Architecture (2007), Massachusetts Institute of Technology" unpublished.K. Elissa, "Title of paper if known," unpublished
6. Mattingly, J.D.: Elements of Gas Turbine Propulsion. McGraw Hill (1998). ISBN 0-07-912196-9
7. http://www.darpa.mil/Our_Work/TTO/Programs/AVM/AVM_Design_Tools_(META).aspx. DARPA META Program report (2011)
8. Sinha, K., de Weck, O.: A network-based structural complexity metric for engineered complex systems. In: Systems Conference (SysCon), 2013 IEEE International, pp. 426–430
9. Weyuker, E.: Evaluating software complexity measures. IEEE Trans. Softw. Eng. **14**(9), 1357–1365 (1988)
10. Stuart, D., Mattikalli, R.: META II: Complexity and Adaptability. The Boeing Company, DARPA META Report 2011
11. Kurzke, J.: Advanced user-friendly gas turbine performance calculations on a personal computer. ASME 95-GT-147 (1995) (GasTurb™)
12. http://www.skybrary.aero/bookshelf/books/2263.pdf
13. Denman, J., Sinha, K., de Weck, O.L.: Technology insertion in turbofan engine and assessment of architectural complexity. In: Eppinger, S.D., Maurer, M., Eben, K., Lindemann, U. (eds.) Invest on Visualization: Proceedings of the 13th International DSM Conference Cambridge, pp. 407–420. Carl Hanser Verlag GmbH & Co. KG (2011)

Empirical Studies in Decision Rule-Based Flexibility Analysis for Complex Systems Design and Management

Michel-Alexandre Cardin, Yixin Jiang and Terence Lim

Abstract This paper presents the results of human subject experiments focusing on the role of decision rules in the study of flexibility and real options analysis (ROA) in design and management of complex engineering systems. Decision rules are heuristics-based triggering mechanisms that help determine the ideal conditions for exercising flexibility in system operations. In contrast to standard ROA based on dynamic programming, decision rules can be parameterized as decision variables, and therefore capture the decision-making process based on specific realizations of the main uncertainty drivers affecting system performance. Similar to standard ROA, a decision rule approach can be used to quantify the benefits of flexibility in early conceptual design studies, and help identifying the best flexible systems design concepts before a more detailed design phase. While many studies demonstrate expected lifecycle performance improvement stemming from a decision-rule based approach as compared to standard design and ROA techniques, very few studies show experimentally their effectiveness in managing flexible engineering systems. This paper presents the results of controlled human-subject experiments involving thirty-two participants evaluating a training procedure in a simulation game environment. The controlled study show that a stochastically optimal flexible strategy combined with an initial policy for the system configuration can improve significantly the expected coverage rate of medical emergencies. These provide insights for further research, development and evaluation of flexible systems design and management strategies for complex engineering systems.

M.-A. Cardin (✉) · Y. Jiang · T. Lim
Department of Industrial and Systems Engineering, National University of Singapore, Block E1A #06-25, 1 Engineering Drive 2, Singapore 117576, Singapore
e-mail: macardin@nus.edu

G. Fanmuy et al. (eds.), *Complex Systems Design & Management*,
DOI 10.1007/978-3-319-49103-5_13

1 Introduction

Complex systems, such as urban infrastructure systems supporting emergency services, energy generation, water and transportations, are inevitably facing a wide range of uncertain factors over their lifecycle (e.g. markets, operational environment, government regulations, technological evolutions, etc.) Standard approaches to systems analysis and design, such as discounted cash flow (DCF) analysis, scenarios planning, sensitivity analysis, etc., typically focus on system concepts and design parameters that aim to optimize system's performance under limited considerations of the main uncertainty drivers affecting their performance. Such approaches often result in fixed-point design alternatives that are rigid in terms of system configuration and planning, and often fail to adapt to unexpected future uncertainties. Since the irreversible capital investments involved in deploying large-scale systems are high, more research is needed to develop systematic procedures that can help both designers and system operators devise and manage engineering systems that are more adaptable in the face of a dynamic environment, with the goal of improving their expected lifecycle performance.

This paper focuses on developing a novel approach based on decision rules to analyze flexibility in complex systems design and management. It proposed an empirical approach to do so, performing human-subject experiments to determine the training procedures that are most likely to improve the expected lifecycle performance of complex systems, with example application in emergency medical services (EMS) systems. Decision rules are heuristics-based triggering mechanisms that help determine the ideal conditions for exercising flexibility in system operations. In contrast to standard ROA techniques, decision rules can be parameterized as decision variables, and capture the decision-making process based on specific realizations of the main uncertainty drivers. They can be used to quantify the benefits of flexibility in early conceptual design studies, and help identifying the best flexible systems design concepts before a more detailed design phase. The proposed decision rules based approach comprises a set of practical heuristic mechanisms that emulate the actual decision making process (e.g. akin to an *if-then-else* logical statement; an example conditional-go decision rule can be *if demand is higher than a certain threshold, expand capacity, else do nothing*). Decision rules can be used to model the behavior of flexible systems, and to study the different phases of the design process. For instance, in the concept generation phase, a flexible system design concept consists of a strategy (e.g. abandonment, switching, capacity expansion, etc.) and enabler (e.g. decision rule). Therefore, concept generation techniques to generate creative decision rules can be studied, as done in [1]. In the design space exploration phase, the optimal values of design and decision rules variables can be found using different types of optimization techniques (e.g. stochastic programing, meta-heuristics, design of experiments methods, robust optimization, etc.). In the process management phase, new training procedures can be developed and their effectiveness tested in an empirical setting to train system operators and decision-makers on how to best use the identified optimal decision rules in operations.

This paper is organized as follows. Section 2 provides further background on flexibility, ROA, and decision rules in the context of engineering systems design and management. It identifies the main research questions and contributions of this study. Section 3 identifies the optimal design and decision rule variables as part of the design space exploration phase, which is then used in the controlled empirical studies. Section 4 presents human-subject studies evaluating the effectiveness of new procedures based on the decision rule method to train system operator on how to best use the optimal decision rules identified in the optimization study. Section 5 presents and discusses the results considering validity and limitations issues. Section 6 concludes and offers directions for future work.

2 Background

2.1 Flexibility in Systems Design and Management

The concept of flexibility is akin to a real option, as it aims to provide system operators with "the right, but not the obligation, to change a system in the face of uncertainty." [2] It captures an emerging paradigm in engineering design whereby designers enable on purpose the system to change pro-actively in the face of uncertainty. Flexibility builds upon ideas and techniques in risk management to support the design of better performing systems, aiming on the one hand to reduce exposure to downside risky scenarios, on the other hand to position it so it can benefit from upside opportunities. Typical flexibility strategies inspired from the standard literature on ROA include staging/phasing capacity deployment, allocating resources dynamically, increasing or shutting down operations as needed, and/or switching between different activity types. Many studies have been conducted so far in different industry sectors, and all have shown significant improvements in terms of expected lifecycle performance, as compared to standard design and project evaluation approaches [3–5].

2.2 Real Options Analysis

The work on ROA focuses on quantifying the economic value of flexibility [2]. For engineering design, standard ROA techniques can be used to evaluate the expected performance of flexible systems design concepts. There are three main approaches to do so: (1) decision analysis, (2) binomial lattice analysis, and (3) simulations. In decision analysis, a simplified version of the more general dynamic programming problem is used to determine the best design decisions at each stage, based on

Bellman's recursive equations [6]. Binomial lattice analysis also relies on a backward induction process. Important differences with decision analysis is that in each stage the uncertainty can either go up or down relative to the previous state, and paths recombine to reduce the number of possible paths and computational complexity. The binomial lattice is essentially a discrete formulation of the Black-Scholes formula [7] developed to value financial options in the 1970s, which provided the first quantitative tools to measure the value of flexibility in financial assets [8]. Under the simulation approach, stochastic scenarios are modeled explicitly, and the performance of the system as governed by the design and decision rule variables is evaluated under each such scenario. Cardin [9] proposed a formal systematic approach to investigate the design space for flexible systems under the decision rule paradigm—see details in Cardin et al. [10].

2.3 Decision Rules

Broadly defined, a decision rule—also referred as implementable policy—is a function that maps the observations of uncertainty data to specific decisions [11]. Decision rules have been developed and used in research on multistage stochastic programming. Garstka, Wets [12] provide a survey of the different types of decision rules available. Four classes exist: zero-order, linear, safety-first and conditional-go. Zero-order decision rules follow a particular optimal plan, no matter how the uncertainty realizes in operations. They are typical outputs from stochastic programming problems whereby the optimal design is found based on a range of possible uncertainty scenarios. Linear decision rules assume that the decision is a linear function of the uncertainty realization, and identifies the best parameters to govern the decision making process [13] They enable dynamic adaptation to changing operating conditions. Safety-first decision rules identify the parameters providing a safety margin, or minimal acceptable performance, in the objective function, when the system is subject to uncertainty realizations. Conditional-go decision rules are adaptable policies that rely on observations of the uncertainty realizations. They are akin to *if-then-else* statements in programming (e.g. take action A *if* uncertainty factor fulfills criterion B, *else* do nothing).

3 Decision Rule Optimization Analysis

An example EMS system design problem is chosen because it provides a wide range of opportunities to study flexibility in engineering systems. In design and planning, EMS systems require thinking about the distribution of the capacity of vehicles and stations used for maintenance and operations along four dimensions: location over two spatial dimensions (e.g. location in the city), site-specific capacity (e.g. number of vehicles maintained and operated at any given station), and time.

Flexibility can be exploited across all dimensions to manage risks and uncertainty. For example, abandonment may involve temporarily closing down a station where incidents are lower than anticipated. System operators may defer construction of new capacity until sufficient needs arise. Capacity can be expanded/contracted for each station depending on needs, or be deployed in phases instead of all at once—which helps deferring costs to the future, and reduce their net present value. If the infrastructure serves different purposes (e.g. fire, ambulance), stations can be designed to accommodate different vehicle types, an example of switching strategy. The flexibility inherent to EMS systems exists in many other examples of urban distributed systems, of which EMS is an example. For instance, deploying and operating liquefied natural gas (LNG) production systems subject to uncertain demand growth, waste-to-energy (WTE) systems facing uncertain growth in waste generation, or real estate projects face similar challenges [14, 15].

3.1 Design Problem

The example engineering system of interest is an urban EMS system providing emergency medical services in Singapore. The design problem is inspired from the one described in [16], albeit modified to suit the needs of this new study. The goal is to determine the optimal siting configuration and ambulance allocation (i.e. the system design) to maximize the expected long-term incident coverage rate, subject to limited resources (e.g., facilities, capital budgets) and uncertainty in incident rates in different geographical locations. The coverage rate is affected by multiple uncertainty sources, ranging from short to long-term, such as incident rates at different times of the day, traffic conditions, construction of new population and industrial estates, overall demographic changes, etc. [17].

3.2 Computer Model

A simulation-based model is developed to assess system coverage rate and other key system performance indicators under different decision sequences and system configurations. The collaborating EMS provider provided the original incident list and requirements for developing the simulation engine, as well as the context for decision-making evaluation (e.g. key performance indicators, typical constraints, etc.) The model was developed in Matlab® using standard techniques in discrete event simulations (DES) and urban operations research [18].

The simulation focuses on dispatching ambulances to handle medical incidents that occur randomly, depending on the spatial configuration of stations, and capacity allocations. The DES engine generates medical incidents randomly—also referred as demand for ambulances. Uncertainty is captured by key parameters, like changing geographical incidents rate, travel times subject to random fluctuations, and variable patient conditions. The description of the simulation engine is divided into two sections: incident generation and incident handling [16]. Incident generation produces a list of new incidents from the list of real incident data. Incident handling is performed during a simulation. The coverage rate cr_n, and operating cost c_n in each scenario n are calculated in Eqs. 1 and 2:

$$cr_n = \frac{FR_n}{IN_n} \times 100 \tag{1}$$

$$c_n = \sum_t \left(\sum_j oc \cdot x_{jtn} + \sum_j mc \cdot w_{jtn} + \sum_i \sum_j q_{ij} \cdot y_{ijtn}\right)(1 + r_t)^{-(t-1)} \tag{2}$$

In Eq. (1), FR_n is the number of incidents responded within 11 min (i.e. time threshold determined by the local EMS provider for a "fast" response), and IN_n is the total number of incidents that occurred over time period T in scenario n. In Eq. (2), oc is the operating cost per unit capacity for a station, mc is the maintenance cost per ambulance, x_{jtn} is the capacity of a station in grid cell j at time t under scenario n, w_{jtn} is the ambulance number in station j at time t, q_{ij} is the assignment cost to assign grid cell i to station j, y_{ijtn} is 1 if grid cell i is assigned to station j at time t and zero otherwise, and r_t is the discount rate representing the opportunity cost of capital. The term $\sum_j oc \cdot x_{jtn}$ is the operating cost of all station at time t, the term $\sum_j mc \cdot w_{jtn}$ calculates the operating cost of all ambulances deployed in period t, the term $\sum_{i=1}^{I} \sum_{j=1}^{J} q_{ij} \times y_{ijtn}$ is the total assignment cost, and the term $(1 + r_t)^{-(t-1)}$ is the discounted factor for the costs.

3.3 Optimization

To apply real options or flexible strategies in the EMS system, decision makers should know when it is appropriate to exercise the flexibilities in operations. A decision rule is just such a triggering mechanism that system operators (or decision-makers) can use to determine when it is appropriate to exercise the real option/flexibility. As explained in Sect. 2.3, conditional-go decision rules are typically based on the observation of a given uncertainty source (e.g. demand, price, technological performance) to which the system is called to adapt. The decision triggers for exercising flexibility in the EMS system regarding station installation and capacity expansion are formulated as conditional-go rules, akin to "*if-then-else*" programming statements, generally structured as follows: $= IF(logic\ test, value\ if\ true, value\ if\ false)$. The statement will give a value "1" if all criteria for exercising the real option condition are satisfied (described in the logical test), meaning that the real option should be exercised. Otherwise, the statement will result in value "0" if any of the criteria is not met, which means that the exercise decision is postponed until new observations are available, or the system lifecycle ends.

There are two types of real options for station deployment: delay or phase capacity deployment, and expand capacity on-site. The "*if-then-else*" statements that parameterize the decision rules in the model are summarized as follows. For the phasing capacity flexibility, "*if the number of incidents in a candidate site j is greater than or equal to de_j within a strategic period s, then a new station with a certain capacity op_j must be installed in candidate site j at $s+1$*". For capacity expansion flexibility, "*if the number of incidents missed in a station within a strategic period s is greater than or equal to me_j, then the current station j must be expanded with a certain capacity oe_j at $s + 1$*". The parameters de_i, op_i, me_j, oe_j are decision rule variables, and the time period t is assumed to be one year. The decision rule for the phasing strategy is that a new upgradable station with a certain capacity op_j in a candidate site j will be built in year $t + 1$ if the number of incidents that occurred at that site in the previous year t is higher than a certain threshold de_j, else no new station will be built. The decision rule for capacity expansion is that an upgradable station j will be expanded with a certain capacity oe_j in year $t + 1$ if it fails to respond to a certain number of incidents me_j in the previous year t, else no expansion will occur.

4 Empirical Studies Using Simulation Games

4.1 Treatments

Four treatment conditions are evaluated in a 2 × 2 design of experiment (DOE) setup, consisting of two factors with two levels each. All material used in experiments (e.g. lecture slides, documents, etc.) are available from the authors upon request. Factor 1 Decision rule (D) consists of decision rule training via a hands-on in-game practice ($D = +1$). Current training ($D = -1$) establishes the baseline when no particular training on the decision rules is provided. The decision rule training ($D = +1$) is devised to help participants acquire sufficient knowledge about flexibility and application of decision rules using the simulation game, without telling them exactly what to do in each round. Step-by-step instructions are provided to guide the participants to observe the evolution of uncertainty, and take actions to build a flexible EMS system according to the recommended decision rules found in the optimization analysis (e.g. demand quantities on different locations). Current decision rule training ($D = -1$) assumes that the participants play the game based on their own background, experience, and the basic training on using the simulation game provided in introduction, without any particular emphasis on uncertainty and flexibility. Flexible decisions are available to all participants, but no guidance is provided on how to make best use of them under the latter treatment.

Factor I captures the decisions on new stations opened (if any) at the beginning of year $t = 1$, which is either provided ($I = +1$) or not provided ($I = -1$). There is no decision rule for station installation and expansion at the beginning of year $t = 1$ since the EMS system design game begins in year 1, and demand information on the previous year cannot be observed in the simulation game. The purpose of including this factor is to account for the impact of initial location of the stations on the expected coverage performance. The flexible decisions will be greatly affected by this initial decision, hence the need to account for it in the analysis. In reality, such initial locations can be found via optimization, or following an existing deployment plan that is already determined by the EMS organization. One treatment ($I = +1$) is therefore given the initial location of new stations to deploy at the beginning of $t = 1$, but not how much capacity to deploy in each location. For treatment ($I = -1$), participants may position new stations and assign capacity at the beginning of $t = 1$ wherever they want.

4.2 Participants

The ideal target population for this study would be the actual stakeholders (e.g. EMS designers, planners, managers, and system operators), if the game were used for training purposes at the EMS provider. Due to the time challenges associated with getting such stakeholders involved in the study, students were recruited on a voluntary basis to participate in these experiments. Thirty-two participants were recruited from university courses in industrial and systems engineering at an institution of higher learning in Asia. They were exposed through their training in operations research to the kind of decisions and tradeoffs that are exploited in the game setup. The population consists of 50 % male/female, 6 % being <25 years old, 81 % between 25–29 years old, and 13 % > 29 years old. Of the sample population, 81 % have <1 year of work experience, 16 % 2–3 years, and 3 % > 3 years of work experience. 91 % of participants already have a bachelor's degree in a related engineering discipline, while 9 % have a master's degree.

4.3 Design and Management Problem

In each round of the game (see Sect. 4.6), tactical and strategic decisions are made. All decisions must be made towards the goal of producing the best coverage rate possible, considering a constrained budget. Strategic decisions include (1) deploying initial capacity of flexible upgradable station (available capacity range from 1 to 3 capacity units), (2) deciding on site location and timing, (3) whether to buy more ambulances, and/or (4) upgrading an upgradable station. Tactical decision involves (5) allocating available ambulances at different sites, depending on site capacity allocations made at the strategic level.

At the beginning of each session, the simulation game is described. Players are given a map of an area of the city used for the game. The map is divided into several grids corresponding to a particular area of the city. Participants' task is explained to design and manage the EMS system by considering the above decisions. Details are provided on the characteristics of stations (i.e. capacity, capital cost, building time), ambulance costs, and on constraints for siting the stations (e.g. one station per grid cell). The objective is explained to maximize the long-term incident coverage rate. Strategic and tactical actions allowed are described. Examples are provided to get familiar with the user interface, such as visualizing incident forecasts in different locations, tracking performance indicators, and some guidance on how to interpret such analytical indicators. Examples are provided to understand how to use the game environment.

4.4 Experiments

A pretest-posttest structure is adopted for each experimental session, so that participants play the game over two sessions, separated by one of the treatment conditions described in Sect. 4.1. Various strategies help mitigate the effects of confounding variables (e.g. human factors, environment variables) likely to influence experimental results. The pretest-posttest structure adds an additional layer of stringency to randomized group assignments to measure potential statistical effects. Randomized assignments to treatment groups dilute group bias due to human factors such as creativity levels, personality, risk aversion level, etc. All presentations are audio-video recorded (e.g. introductions to the design problem and game interface) to control for information variability, and the same document is used for flexibility training. The same time is allocated in each session to account for the fact that participants may get more familiar with the game over time, which will affect lifecycle performance and scores. Also to prevent participants' getting too familiar with the incident scenarios, a different incident list is used in sessions 1 and 2. All games are subject to the same assumptions for the parameters and decision variables, providing a higher degree of measurement reliability.

4.5 Data

Raw data is collected via the computer engine, and online survey collection is done via Qualtrics® for demographic information and qualitative user impressions. The computer model records the impact of different design and management decisions made in each round based on the relevant performance indicators, and saved to standard ASCII text format. For example, final coverage rate at the end of 10 rounds is recorded for each participant as a lifecycle performance metric. Survey results are recorded as the sum of Likert scores for all constructs.

4.6 Simulation Game

The computer model underlying the simulation game is the one described in Sects. 3.1 and 3.2. The simulation game interface is developed as an overlay to the DES model, so that players can make decisions under different treatment conditions, several rounds run can capture the effects on performance assuming the accelerated passage of time, and one can therefore study the impact of different decisions under different treatment (or training) conditions. This game consists of 10 rounds where each round corresponds to 1 year of real-world operations. The stated objective is to maximize incident coverage rate over 10 years subject to a constrained budget, which captures important tradeoffs in terms of EMS service quality, and financial resources (i.e. if unlimited resources were available, the best policy would be to deploy stations with maximum capacity in each sector).

Decisions available to participants aim to capture realistic trade-offs in EMS management and resource allocation: (1) service quality vs. budget limit, (2) cost savings from centralizing capacity and associated economies of scale versus wider coverage and shorter response distance, (3) deferring investment and using more information to make decisions vs. early commitment of fund providing greater utility of resources, and (4) greater flexibility vs. cost of acquiring the flexibility. There is no limit to the number of decisions players can make each round. Each decision, however, is constrained by the overall budget limit.

The graphical user interface (GUI) overlaying the DES engine is developed in Matlab®. It shows a map of a particular sector of Singapore, divided into 8×13 grid cells. Participants can right-click on any grid cell to make strategic decisions (e.g. phasing station installation, upgrading a flexible station, purchasing new ambulances), and left click to make tactical decisions (e.g. allocating or moving ambulances to different sites). The GUI also shows the round number, total available budget for all rounds (e.g. $28.66 million at the beginning of the game), current coverage rate (e.g. $cr = 0$), and short instructions for game playing. Other tabs show alternative visualization tools, and give access to other settings (e.g. incident history and forecasts, performance indicators).

5 Results and Discussion

Tables 1 and 2 show the optimal solutions obtained for the rigid and flexible designs, respectively, using simulation-based optimization. The rigid solution follows a fixed deployment plan, based on expected incident patterns in the future. The flexible solution is more dynamic.

For instance in Table 2, no facility should be deployed at the beginning of time $t = 1$ (i.e. $io_1 = 0$). If the number of incidents in the area surrounding site 1 is beyond threshold $\mathrm{de}_1 = 7795$, the phasing flexibility is triggered, and $op_1 = 1$ unit of capacity should be deployed. Then, each round where the number of lost

Table 1 Optimal solution for rigid design

Candidate site to build at the beginning of the first strategic period	3	4	6	7	8
Capacity	1 unit	2 unit	3 unit	1 unit	1 unit

Table 2 Optimal flexible solution with decision rules

Variable	Candidate site								
	1	2	3	4	6	7	8	9	10
de_j	7795	2630	1700	2782	482	214	1571	4505	6437
op_j	1	1	1	2	2	1	1	1	1
me	1800	1800	1800	1800	1800	1800	1800	1800	1800
oe_j	1	1	1	1	1	1	1	1	1
io_j	0	0	0	1	1	1	0	0	0

incidents reaches $me = 1800$ in the surrounding environment, an open station should deploy additional capacity of $oe_1 = 1$ unit. A similar approach is used to manage dynamically each station j = 1, 2, …, 10. The performance improvement in the flexible system stems from the fact that the system is better able to deploy the capacity only *if and when it is needed*, thereby making better use of available resources (i.e. both in terms of facilities, emergency vehicles, and money).

5.1 Experimental Study

Before conducting the human subject experiment and reporting the results, a computational study is performed to evaluate the improvement in total coverage rate (*CR*) under different treatment conditions. In the computational experiments, decisions to design and manage the flexible system are automatically generated by the computer to simulate the possible decisions made by human under the four treatment conditions, using the same pretest-posttest structure. Two distinct sample scenarios are used for sessions 1 and 2, and ten replicates are obtained in each treatment group. Responses ΔCR captures the possible improvement on quantitative lifecycle performance measurement on total coverage rate. Under treatment condition $D = -1$, the computer selects a combination of decision variables randomly, where each decision variable is equally likely (assumed uniform distribution), while under $D = +1$. Under condition $I = -1$, station location and capacity at beginning of period $t = 1$ are not provided. Station location is provided under condition $I = +1$, but not the capacity. In essence, the computer experiment gives a sense of possible extremes: one the one hand supposing humans implement the decision rule perfectly under case $D = +1$, on the other hand assuming no inherent logic or experience under condition $D = -1$, and only random decision-making constrained by the design space defined.

Equation 3 shows the GLM of the main and interaction effects. Results from the regression analysis shows that both factors D and I have statistically significant effects on the response ($\beta_D = 0.165$, $p_D = 0.00$; $\beta_I = 0.158$, $p_I = 0.00$). As compared to the baseline conditions $D = -1$ and $I = -1$, both decision rule $D = +1$ and station location $I = +1$ at the beginning of period $t = 1$ generate statistically significant performance improvements.

Equation 4 shows the GLM of the main and interaction effects obtained for the experiments with human subjects, considering eight replicates (or participants) in each treatment group. It shows that both D and I also have statistically significant main effects on the response ($\beta_D = 0.078$, $p_D = 0.01$; $\beta_I = 0.067$, $p_I = 0.03$), which is consistent with the results from the computational experiments.

$$\Delta CR = 0.294 + 0.165D + 0.158I + 0.036DI \tag{3}$$

$$\Delta CR = 0.156 + 0.078D + 0.067I - 0.058DI \tag{4}$$

A combination of decision rules and initial location procedure training produces the best lifecycle improvement in coverage rate among all four treatments. Observing both statistically significant main effect shows that decision rule training has a main effect, independent of initial location, and vice versa. Both main effects are weaker with human subjects, although still statistically significant. This is because humans may not always implement the flexible rules perfectly in experiments, and/or rely on self-made strategies when no decision rule is provided, instead of completely random assignments as done in computer experiments.

6 Conclusion

This paper presents the results of controlled computer-based and human subject empirical studies evaluating the effectiveness of a decision-rule based approach to support the design and management of flexible engineering systems. A decision-rule approach to flexibility analysis contrasts and compares to standard ROA techniques based on dynamic programming by focusing on a parameterization of the key decision variables, and therefore emulating the decision-making process based on specific realizations of the main uncertainty drivers affecting system performance. This parameterization helps compute the best decision variables to design (i.e. physical initial design) and manage (i.e. decision rules) flexible systems in operations using optimization techniques, with the goal of improving expected lifecycle performance in the face of uncertainty.

In an example application in emergency medical services (EMS), the studies show how a decision rule can be conceptualized, parameterized, and used to conduct novel research in two important phases of the process for enabling flexibility in complex systems: design space exploration, and process management [19]. Controlled computer and human subject experiments are conducted to evaluate the effectiveness of new procedures to train system operators to make best use of the decision rules identified computationally.

The empirical study with human subjects show statistically significant main effects of both decision rule training and initial location decisions on expected lifecycle performance (ΔCR) using a simulation game environment. These results are new and important: they show that it is possible to identify a stochastically optimal design solution using a computer-based model of a complex system, train system operators on how to use it using a short and effective training, and demonstrably improve the expected lifecycle performance of the system. The results demonstrate the effectiveness of the decision rule technique to improve the system performance in the face of long-term uncertainties.

Acknowledgments This study was funded by the National University of Singapore (NUS) Faculty Research Committee via MoE AcRF Tier 1 grant WBS R-266-000-079-112. The authors thank Howard Ka-Ho Yue and Zhang Sizhe for their inputs on the modeling framework. We also thank all students who actively participated in the study.

Appendix

List of important parameters and variables

Notation	Value	Unit	Definition
S	10	Year	Number of strategic periods of the system lifecycle
T	40	Quarter	Number of tactical periods of the system lifecycle
N	5	–	Number of scenarios considered
I	10	–	Number of demand grid cells considered
J	10	–	Number of candidate sites for station installation
sc_l	2.00, 3.73, 5.38	$, Million	Installation cost for upgradable station with $l(=1,2,3)$ capacity
uc_l	1.80, 3.36, 4.84	$, Million	Upgrading cost for expanding a station with $l(=1,2,3)$ capacity
oc	0.01	$, Million	Operation cost per unit capacity for a station
ac	0.1	$, Million	Unit cost per ambulance
mc	0.01	$, Million	Maintenance cost per ambulance
r_s	12 %	–	Annual discount rate
r_t	2.87 %	–	Quarterly discount rate

(continued)

(continued)

Notation	Value	Unit	Definition
x_{jtn}	[0, 4]	–	The capacity of the station on station j at t under scenario n
w_{jtn}	[0, 4]	–	Number of vehicles allocated on station j at t under scenario n
y_{ijtn}	1 or 0	–	1 if grid cell i is assigned to station j at t under scenario n, and 0 otherwise
de_j	[1, 8000]	–	Threshold of incident number occurred on grid j for triggering the installation of a new station at $s=2, \ldots, S$
d_{itn}	[1, 8000]	–	The number of emergency incidents in location i within tactical period t under scenario n
o_{jsn}	1 or 0	–	1 if total number of incidents occurred on grid j over a strategic period s − 1 is greater than or equal to de_j at $s=2, \ldots, S$, and 0 otherwise
u_{jsn}	1 or 0	–	1 if the total number of incidents missed in station j over a strategic period s − 1 is greater than or equal to me_j at time s = 2,…,S, and 0 otherwise
op_j	[1, 3]	–	Optimal capacity to be deployed if new station is opened on j at $s=2, \ldots, S$. If new station is not opened on j, $op_j=0$
io_j	[0, 1]	–	$io_j=1$ if new station is opened on j at $s = 1$
op_1_j	[1, 3]	–	Optimal capacity to be deployed if new station is opened on j at beginning of time $s = 1$.
oe_j	[1, 3]	–	Optimal amount of capacity to be expanded on j if a station on j is expanded.
me	[1, 3200]	–	Optimal amount of lost incidents in a station to trigger the flexibility of capacity expansion at $s=2, \ldots, S$
ω_{jtn}	[1, 8000]	–	The amount of lost incidents at station j within tactical period t under scenario n
M_1, M_2, M_3, M_4	100000	–	Constants whose values is large, for optimization purposes
U_0	3200	–	The upper bound for variable me
L_0	0	0	The lower bounds for variable me
δ_1	1600	–	The iterative value for searching the optimal value of me

References

1. Cardin, M.-A., Kolfschoten, G.L., Frey, D.D., Neufville, R., Weck, O.L., Geltner, D.M.: Empirical evaluation of procedures to generate flexibility in engineering systems and improve lifecycle performance. Res. Eng. Des. **24**(3), 277–295 (2013). doi:10.1007/s00163-012-0145-x
2. Trigeorgis, L.: Real Options. MIT Press, Cambridge (1996)
3. de Weck, O.L., de Neufville, R., Chaize, M.: Staged deployment of communications satellite constellations in low Earth orbit. J. Aerosp. Comput. Inf. Commun. **1**, 119–136 (2004)

4. Koh, E.C.Y., Caldwell, N.H.M., Clarkson, P.J.: A method to assess the effects of engineering change propagation. Res. Eng. Des. **23**(4), 329–351 (2012). doi:10.1007/s00163-012-0131-3
5. Mikaelian, T., Rhodes, D.H., Nightingale, D.J., Hastings, D.E.: A logical approach to real options identification with application to UAV systems. IEEE Trans. Syst. Man Cybern. Part A **42**(1), 32–47 (2012)
6. Bellman, R.: On the theory of dynamic programming. In: Paper Presented at the Proceedings of the National Academy of Sciences of the United States of America, Santa Monica, CA, United States (1952)
7. Black, F., Scholes, M.: The pricing of options and corporate liabilities. J. Polit. Econ. **81**(3), 637–654 (1973)
8. Cox, J.C., Ross, S.A., Rubinstein, M.: Options pricing: a simplified approach. J. Financ. Econ. **7**(3), 229–263 (1979)
9. Cardin, M.-A.: Facing reality: design and management of flexible engineering systems. Master of Science Thesis in Technology and Policy, Massachusetts Institute of Technology, Cambridge, MA, United States (2007)
10. Cardin, M.-A., de Neufville, R., Geltner, D.M.: Design catalogs: a practical approach to design and value flexibility in engineering systems. Syst. Eng. **18**(5), 453–471 (2015). doi:10.1002/sys.21323
11. Shapiro, A., Dentcheva, D., Ruszczyński, A.P.: Lectures on Stochastic Programming: Modeling and Theory, vol. 9. SIAM (2009)
12. Garstka, S.J., Wets, R.J.-B.: On decision rules in stochastic programming. Math. Program. **7**(1), 117–143 (1974)
13. Kuhn, D., Parpas, P., Rustem, B.: Bound-based decision rules in multistage stochastic programming. Kybernetika **44**(2), 134–150 (2008)
14. Cardin, M.-A., Ranjbar Bourani, M., de Neufville, R.: Improving the lifecycle performance of engineering projects with flexible strategies: example of on-shore LNG production design. Syst. Eng. **18**(3), 253–268 (2015). doi:10.1002/sys.21301
15. Cardin, M.-A., Hu, J.: Analyzing the tradeoffs between economies of scale, time-value of money, and flexibility in design under uncertainty: study of centralized vs. decentralized waste-to-energy systems. ASME J. Mech. Des. **138**(1), 011401-011401-011411 (2016). doi:10.1115/1.4031422
16. Cardin, M.-A., Jiang, Y., Yue, H.K.H., Fu, H.: Training design and management of flexible engineering systems: an empirical study using simulation games. IEEE Trans. Syst. Man Cybern. Syst. **45**(9), 1268–1280 (2015). doi:10.1109/TSMC.2015.2392072
17. Ong, M.E.H., Ng, F.S.P., Overton, J., Yap, S., Andresen, D., Yong, D.K.L., Lim, S.H., Anantharaman, V.: Geographic-time distribution of ambulance calls in singapore: utility of geographic information system in ambulance deployment (CARE 3). Ann. Acad. Med. **38**(3), 184–191 (2009)
18. Larson, R.C., Odoni, A.R.: Urban Operations Research. Prentice-Hall, NJ, United States (1981)
19. Cardin, M.-A.: Enabling flexibility in engineering systems: a taxonomy of procedures and a design framework. ASME J. Mech. Des. **136**(1), 1–14 (2014). doi:10.1115/1.4025704

Requirements Quality Analysis: A Successful Case Study in the Industry

Elena Gallego, Hugo-Guillermo Chalé-Góngora, Juan Llorens, José Fuentes, José Álvarez, Gonzalo Génova and Anabel Fraga

Abstract This case study analyses the applicability of a Quality Improvement Process that will enhance the quality of the requirements using an existing requirement specification to seed the Knowledge Base with the organization know–how. This Knowledge Base drives the quality assessment and directs the requirement authors to the areas that can be improved. The updated specification feeds back into the Knowledge Base thereby institutionalizing continuous process improvement into Alstom. The case study has been developed by means of a Proof of Concepts using the RQS suit tools to gather the knowledge (KM), analyse the quality (RQA) and authoring requirements (RAT).

E. Gallego (✉) · J. Llorens · J. Álvarez (✉) · G. Génova (✉) · A. Fraga (✉)
Universidad Carlos III de Madrid, LEGATEC Technology Park. 16,
Margarita Salas St. 2nd Floor, 28919 Leganés, Madrid, Spain
e-mail: elena.gallego@kr.inf.uc3m.es

J. Álvarez
e-mail: jose.alvarez@kr.inf.uc3m.es

G. Génova
e-mail: ggenova@kr.inf.uc3m.es

A. Fraga
e-mail: afraga@kr.inf.uc3m.es

J. Llorens
e-mail: llorens@kr.inf.uc3m.es

H.-G. Chalé-Góngora (✉)
ALSTOM, 48 rue Albert Dhalenne, 93482 Saint-Ouen Cedex, France
e-mail: hugo-guillermo.chale-gongora@transport.alstom.com

J. Fuentes (✉)
The Reuse Company, LEGATEC Technology Park. 16, Margarita Salas St.
2nd Floor, 28919 Leganés, Madrid, Spain
e-mail: jose.fuentes@reusecompany.com

G. Fanmuy et al. (eds.), *Complex Systems Design & Management*,
DOI 10.1007/978-3-319-49103-5_14

1 Introduction

In order to assure that requirements properly fulfill with stakeholders needs, and they are validated under real circumstances, producing requirements quality analysis and management seems to be an excellent approach to minimize the failure ratio in the requirements engineering process. As can be seen in Fig. 1 the costs to extract defects along the production lifecycle greatly increase. The longer it takes to detect the failure, the more expensive to solve it.

As costs to extract defects highly increase along the stages of the project, half of the defects could be solved in the requirements definition stage. As shown in Fig. 2, over half of defects are attributed to requirements problems and over 80 % of rework effort is spent on requirements related defects.

Figure 3 shows the main results of a study made in the framework of the RAMP (Requirements Analysis and Modelling Process) project within AFIS (French Association on Systems Engineering). These results show that Requirements Quality remains a challenge to achieve better performance within projects. The

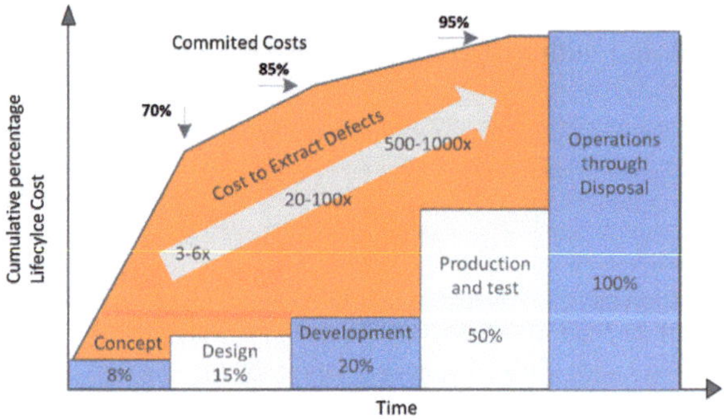

Fig. 1 Costs to extract defects along for the different stages of the project [1]

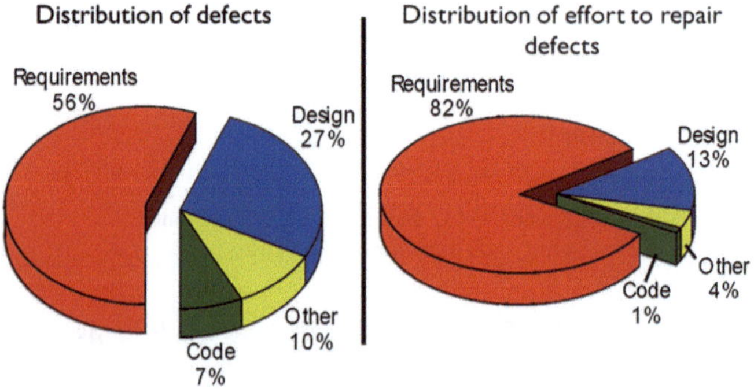

Fig. 2 Distribution of defects and efforts to repair defects [2]

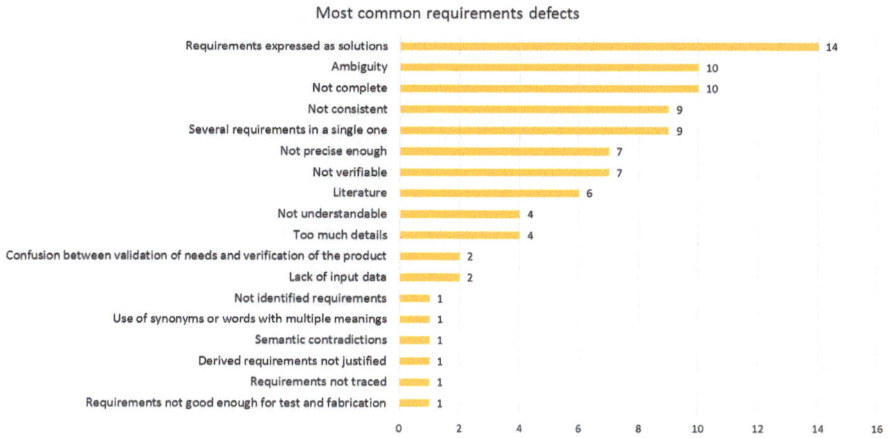

Fig. 3 Most common requirements defects. *Source* Gauthier Fanmuy—the RAMP project:—AFIS

most frequent faults are, express requirements as solutions, ambiguous requirements, not consistent, not complete, several needs in a single requirement, imprecision or unverifiable requirements, among others.

The PoC application within every industry domain covers the main processes to design and elaborate requirements with the adequate quality, gives recommendations to write correct requirements, and defines relevant requirements structures for specific requirement types used in the organization. These outcomes are applicable to anyone writing requirements within the organization, those who perform requirements engineering activities, anyone interested on requirements engineering improvement, and those who are involved in the requirements engineering process for any system in the organization.

2 Proof of Concepts in Context

The Proof of Concepts, PoC, as it is applied in this case study, has been designed to configure a customized methodology for implementing a continuous quality improvement process in a given organization.

During the deployment of a PoC, the main elements to define are the tools to support the continuous quality improvement process and an analysis of the Verification & Validation, V&V, activities within the organization's requirements management and development processes.

The main activities during a PoC are:

- Complete analysis of the quality: the tools perform an analysis of single requirements and sets of requirements covering the three main quality typologies, Correctness, Completeness and Consistency, CCC.
- Quality Control Process: Definition of a process to formalize the requirements quality control, following the Requirements V&V methodology.
- Management Plan: Design of a customized requirements quality management plan to achieve the quality priorities.
- Knowledge Base Design: Knowledge Manager(s) or Domain Engineer(s) design and manage the knowledge base(s) of the organization.
- Requirements Authoring: Requirements Authoring as a mean to improve requirements' quality computer-aided by means of a Requirements Authoring Tool for requirements development.

All the activities carried on during the PoC deployment are directly related to Requirements Engineering main activities.

3 Requirements Engineering: The Vee Model

Requirements Engineering allows both supplier and acquirer of a determined system to develop and understand the requirements set that complies with the system or stakeholders needs. Requirements Engineering is concerned with discovering, eliciting, developing, analyzing, determining verification methods, validating, communicating, documenting and managing requirements [3]. As result of the Requirements Engineering process, it is possible to manage the complexity of the system using a decomposition and integration paradigm by means of the V-Model. As shown in Fig. 4, the left side of the "V" cares about doing the right thing by requirements decomposition, while the right side cares about doing the right thing right by the system (at any level) validation and verification.

To properly assure the lack of potential failures in the system, there is a defined process to determine the requirements engineering sub activities, which covers specific verification and validation activities for requirements. As shown in Fig. 5, it is possible to define the set of sub activities to assure requirements V&V. Before starting, it is important to properly define the scope of both requirements verification and validation:

- Requirements Verification: whether the specified requirements are free from internal errors.
- Requirements Validation: whether the specified requirements properly express stakeholders needs. This activity must be accomplished within the Requirements Engineering process, before the requirements specification is delivered to the development group.

The kind of errors that can be detected and eliminated from requirements prior to validation, include:

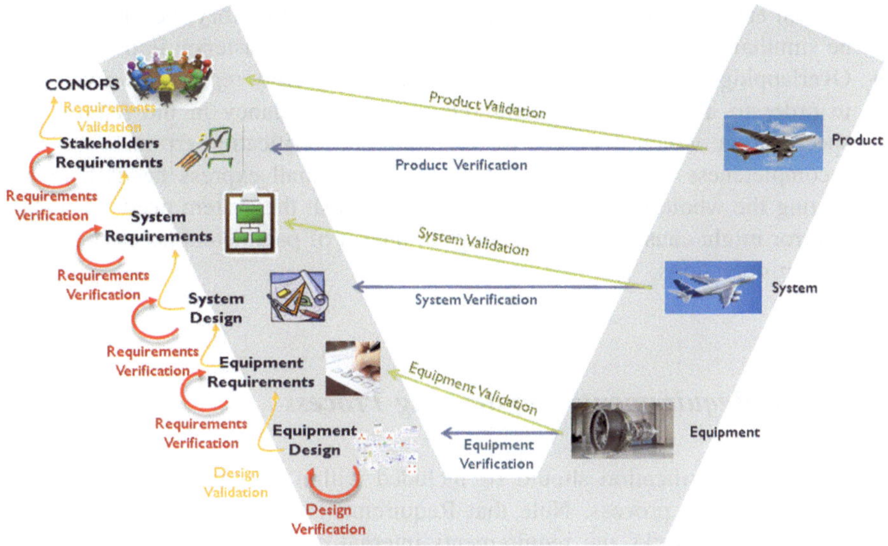

Fig. 4 The V-Model: Integration and decomposition paradigm

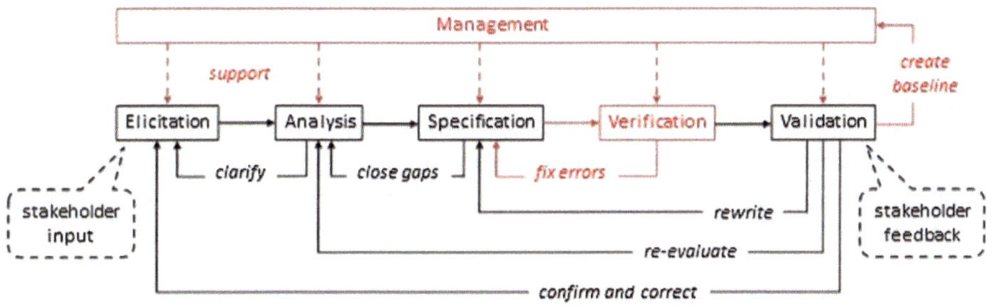

Fig. 5 Requirements sub activities completed with requirements verification and management [4]

- Semi-formal errors: some aspects of correctness that are not strictly logic or formal, such as:
  - Vocabulary errors: requirements must adopt a consistent domain vocabulary throughout the specification, avoiding the use of synonyms and inconsistent, vague or ill-defined terms.
  - Grammar errors: requirements must follow the grammatical rules of the natural language they are written in; by extension, too long or ill-structured sentences can be also detected and corrected.
  - Structural errors: define more or less rigid textual patterns to write requirements, in order to achieve some desirable properties of requirements (completeness, unambiguity, precision, atomicity, etc.)

- Formal errors: different requirements might be contradictory, i.e. impossible to be simultaneously satisfied. They mostly apply to consistency issues.
- Overlapping errors: every requirement shall exclusively represent a given need, in order to avoid possible contradictions or redundancy in the requirements specification, which may lead to validation and verification errors.
- Incompleteness errors: the set of requirements shall express needs by representing the whole necessary factors to accomplish the system needs. This kind of error might cause a lack of control in the set of parameters necessary for the system operations.

3.1 The Requirements Engineering Process

Requirements Verification should be included within a comprehensive Requirements Engineering process. Note that Requirements Verification, as it has been defined in here, checks the requirements internal consistency, properties around them, guidelines, etc. Figure 5 shows the process of Requirements Verification to include the debugging of semi-formal internal errors (Fig. 6).

The processes and sub-activities carried on this proof of concepts tries to be aligned with the previously described approach to Requirements V&V. In order to generate proper Requirements Specifications, the requirements must be written and controlled in a correct, consistent and complete way, and they must be checked to be so, by the requirements verification activity.

This activity must be performed by means of the analysis and management of the requirements quality, whether it is analyzed individually or to the set of requirements in the specification.

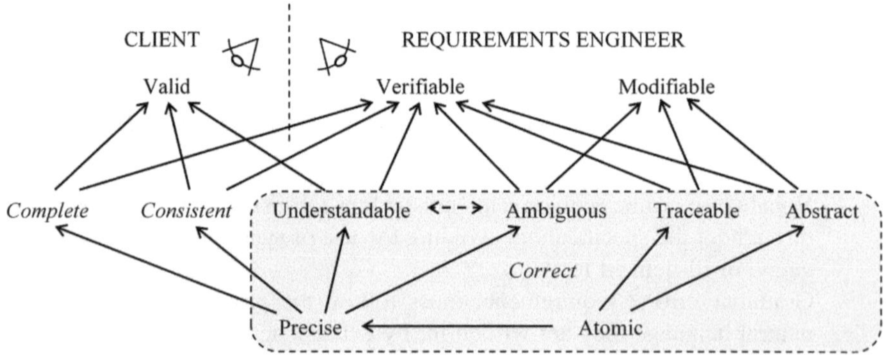

Fig. 6 The Correctness, Consistency and Completeness (CCC) approach for the different properties (*Source* G. Genova, J.M. Fuentes, J. Llorens, and O. Hurtado. A framework to measure and improve the quality of textual requirements: Requirements Engineering Journal. DOI 10.1007/s00766-011-0134-z)

4 Managing Requirements Quality

Along the Requirements Engineering process, the Requirements Quality Analysis shall be applied in order to successfully achieve the main goals of the product development cycle.

Two main ways to define the requirements quality analysis can be considered. From the point of view of the customer, the quality of the requirements shall be such that it needs are validated. From the point of view of the requirements engineer, the two main properties to be supported are to ensure that requirements are verifiable and modifiable. These main properties can also be split into complete, consistent and correct characteristics. The last one, correctness, is a group of several properties that only apply to individual requirements instead of the set of requirements and may be interconnected by their purpose.

4.1 Properties Covered by the REUSE Company's CCC Approach

From the point of view of the correctness, during the PoC, there has been defined a set of metrics to improve the analysis of requirements, by finding defects for every individual requirement statements for a given specification. There are different possibilities to analyze defects based on properties typology:

- Based on the RMS
- Based on the text
- Based on list of sentences
- Based on the domain
- Based on the verbal mode and voice
- Based on pattern matching

The consistency and completeness characteristics, unlike correctness, are applied to the set of requirements instead of being focused on the induvial requirement statement. During the first stages of every PoC, efforts are centered primarily on the correctness metrics, due to the selected maturity levels policy applied in most of organizations to improve the quality of requirements. More details about this topic are defined in the *Quality Management Definition: Maturity lifecycle* section of this document.

The aim of the completeness metrics is to improve the quality analysis by identifying defects of the specification when it does not represent a complete definition of the product.

The consistency approach analyzes the absence of conflict among a set of requirements. Consistency metrics applied to a set of requirements evaluate the consistency of the specification by analyzing if each need in form of requirement is expressed in only one requirement, and if the different requirements do not conflict with each other in any form.

4.2 Quality Characteristics to Measure

Individual requirement statements should take into account a set of characteristics to ensure that the requirement is properly written and structured. In order to consider those characteristics, a set of different metrics can be assigned to each one of them. If the metrics assigned to a particular characteristic produce good values, it could be possible to consider that a single requirement or a requirements specification do not present problems for it.

The relevant characteristics considered in the PoC methodology for analyzing the quality of requirements, either individually or as a set are defined in the INCOSE Guide for Writing Requirements [1], and must be considered for producing a good requirements specification, even if no metrics can be assigned to them:

- Appropriate
- Unambiguous
- Complete
- Singular
- Verifiable
- Correct
- Conforming
- Completeness of the Specification

For every characteristic are assign different metrics, whether it comes from the INCOSE Guide for Writing Requirements [1], the International Standard ISO/IEC/IEEE FDIS 29148 [3], or from other sources (The REUSE Company), which can be used to perceive how it globally affects the requirements.

Each metric purpose is to measure the coverage of the different characteristics along the evolution of the specification's quality. Resulting from the PoC analysis, every metric provides a description, a set of recommendations with examples, and the quality functions that should be applied to each metric when analyzing the quality.

4.3 Quality Management Definition: Maturity Lifecycle

To write requirements in a way that can be consider correct, consistent and complete is not, in most cases, a straightforward task. Usually, as always happens when dealing with quality issues, it is an incremental activity, where the persons involved in these kind of tasks must gain skills, capabilities and abilities. Within the quality domain, this incremental view is the kernel of the quality management process. A Plan-Do-Check-Act approach, called PDCA, is a common pattern.

The maturity lifecycle process designed during the PoC deployment, defines a requirements quality model based on quality levels. In an incremental way, every

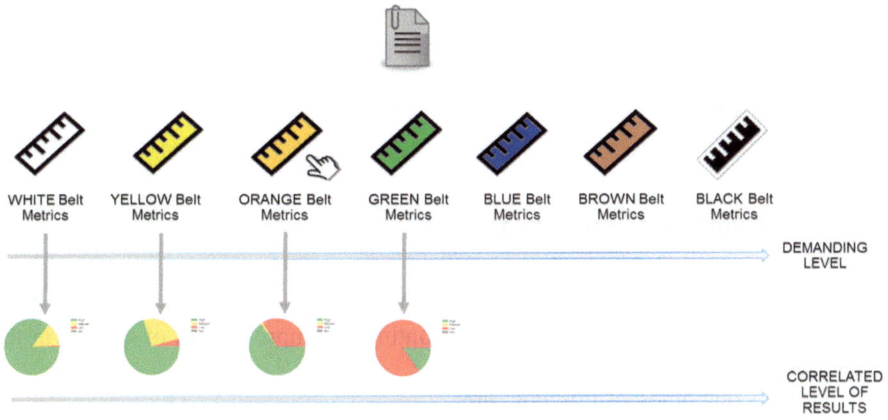

Fig. 7 Requirements quality improvement for the different levels

organization, department or team traverses from its initial level (whichever it is) towards the most mature level, in an endless process (Quality Improvement shall never end). In every case, the quality level must be selected by the corresponding team, for a particular project, and the requirements specification quality is measured applying the quality configuration of the selected level.

The approach applied in the scope of the PoC defines a set of quality maturity levels, which cover the different characteristics progressively for the learning curve in the Requirements Quality Management. There may exist different levels of experience around requirements writing inside an organization, where they can depend on the engineer(s) or the department that is working on a given project. By applying the quality maturity levels approach, it is possible to effectively evolve in the quality process with a correct and incremental distribution of the efforts.

These maturity levels are represented following the nomenclature of the colored obis applied in the martial arts (White for the less mature and Black for the most mature). Figure 7 intends to show the expected quality evolution for the same set of requirements along the first four colored belts, without applying any modification into the specification. As every maturity level demands and analyzes more quality issues, a given specification should have a lower overall quality level when a more mature level is applied. In this sense, it is possible to see the correlation between the defined levels and the expected results.

The organizations should traverse towards reaching the most mature level, and once in this level, they should control and improve the quality of a particular specification to get an accepted level.

Finally, once a maturity level is reached, to select a more mature level implies that all the quality metrics that are controlled at the previous level become compulsory for the next one. This way of operating assures that quality from low maturity levels must always be controlled at more mature levels.

In the scope of the PoC deployment the definition of each level has been carefully designed to cover the main characteristics with a determined effort, in order to be more efficient to plan the strategy of the Requirements Quality Management process.

5 The Proof of Concepts Process

The Proof-of-Concepts follows the PDCA process (Plan—Do—Check—Act), planned as the way to proceed with the continuous improvement of quality in the organization's requirements specifications.

The main outcomes of this PoC are:

1. The definition and agreement of a set of quality metrics according to the organization guidelines and the current state of the art.
2. The configuration of these rules in the Requirements Quality Suite tool.
3. The definition of a first ontology describing the business knowledge of the organization (for one particular domain).

The PoC process is aimed to improve the requirements quality, as shown in Fig. 8, and covers the sub-processes from an initial requirements specification, which is going to be analyzed and improved, to the same requirements specification improved according to a developed ontology that represents the business knowledge of the Organization.

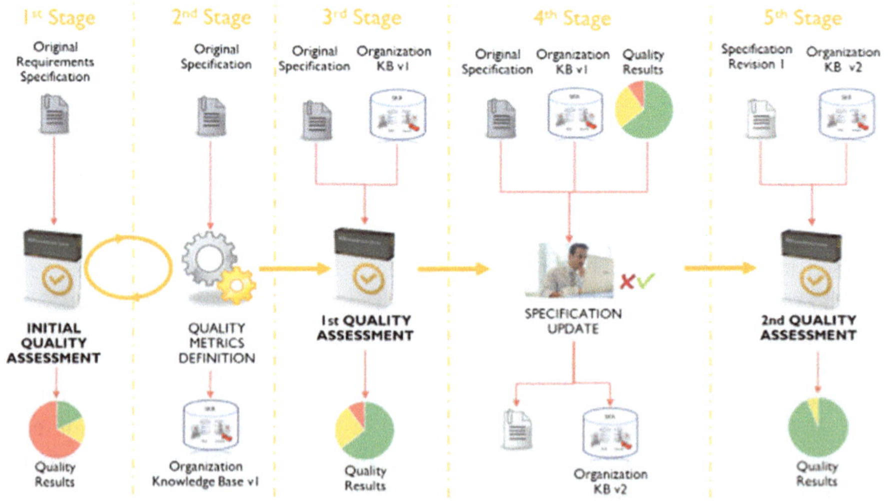

Fig. 8 Requirements quality improvement process for the PoC development

5.1 The PoC Sub-processes

When the PoC is defined according to the approach defined in this document, it is possible to distinguish four main sub-processes or stages in the requirements quality improvement cycle from any specification to a consecutive better quality level.

5.1.1 Stage #1

The assets needed to begin with the quality improvement process are the original requirements specification that is being improved in the PoC, and the ontology provided by The Reuse Company, which supports the essentials to start building the specific knowledge base for the organization in a given domain.

Once the assets has been chosen, it is time to analyze the requirement specification with the RQA tool by assessing the metrics provided by TRC in the default ontology, in order to know the initial quality status of the Requirements Specification according to the out-of-the-box set of metrics. These metrics have been carefully defined based on the INCOSE Guide for writing requirements and the TRC experience.

The first quality assessment results suggest how should be configured the set of metrics for each quality level (colored belts) in the evolutionary approach for the requirements quality improvement. Since the engineer starts to analyze the original requirements specification, an iterative process of evaluation, metrics configuration and ontology population starts until the optimal configuration is defined. The evaluation indicates whether the user has properly defined the proposed sets of metrics, or is necessary to improve the belts definition and the organizational knowledge base.

This sub-process is specifically defined in Fig. 8, the actions "metrics belts definition" and "First ontology elaboration".

The result of the "stage #1" sub-process is the first version of the organization knowledge base, in which the metric belts have been defined and the main knowledge base elements have been included.

5.1.2 Stage #2

The second stage of the process is focused on the evaluation of the actions overtaken in the previous stage. The original requirements specification, together with the first version of the organizational knowledge base resulting from the first stage, the engineer executes the first quality assessment for the predefined belts.

For each defined metric belt, the user will assess the quality metrics to the requirements specification. From every assessment, a quality report will be generated in order to analyze how well suits the specification with the created ontology and the defined quality belts.

The result of this stage is a collection of requirements quality reports generated from the quality assessment in RQA, which will aid the user in the next stage.

5.1.3 Stage #3

In the third stage, the goal is to get an improved requirements specification, from the original one based on the first organization ontology, and an improved organization knowledge base, which will be improved according to the needs raised in the improvement process of the requirements specification.

Once the requirements specification and the ontology have been selected in the RQA, the engineer improves every issue in the requirements, detected by the quality analysis made in the previous stage with RQA.

5.1.4 Stage #4

The last stage of the process will result in an improved specification from the first defined quality level, for instance the white belt, to the following quality level, yellow belt.

The improvement process finish when the requirement specification does not have any metric that contributes with low quality to the requirements. If in this stage, there still are metrics contributing low quality, the engineer should go back to the previous stage (stage #3) to continue improving the requirements quality.

In this last stage, the result is a set of quality reports that verify the quality of the requirements for the selected quality metrics belt.

6 Results

During the Proof of Concepts deployed in Alstom to test the capability of the approach, we ended with an improved requirements specification complying with the specified quality priorities, a defined process and methodology to improve the quality of requirements, a knowledge base with all the information around requirements engineering in the organization, as glossaries, breakdown structures and patterns.

In Figs. 9 and 10 there are shown the results regarding the knowledge base evolution, achieved with the application of the PoC. The PoC started analyzing the quality of a given specification according to the metrics out-of-the-box provided by the RQS suit installers. After analyzing the results and define the quality improvement process, the knowledge base ended with 2,195 new terms to be included into the glossary in the knowledge base, 955 new relationships or terms to be included into the System Conceptual Model, 70 new patterns and 26 different pattern groups.

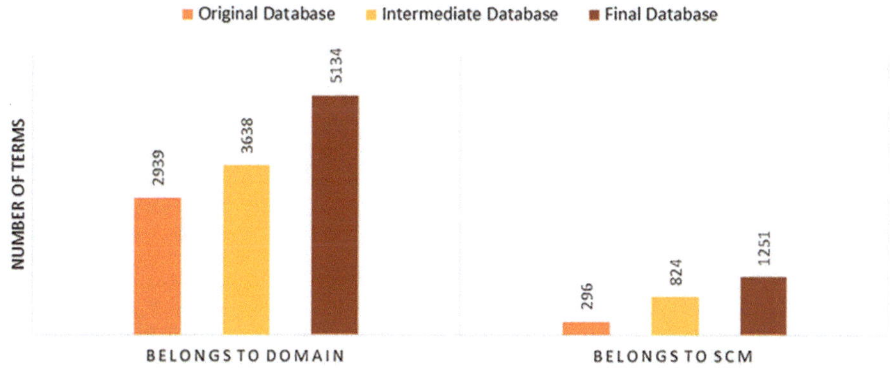

Fig. 9 Terminology evolution from the initial state of the ontology to the first analysis results

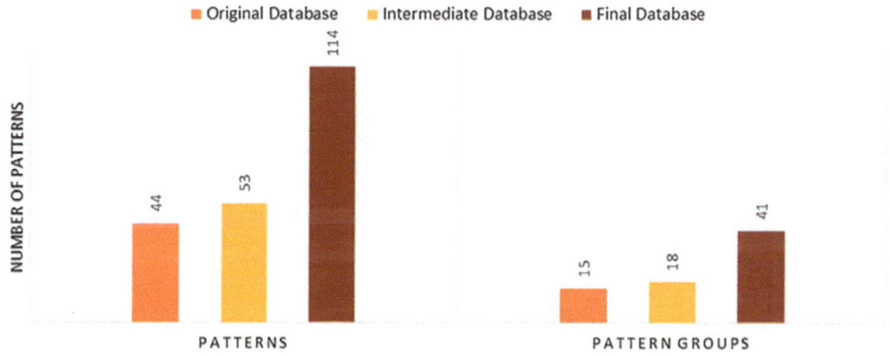

Fig. 10 Patterns evolution from the initial state of the ontology to the first analysis results

Once the first version of the knowledge base was built, and the quality improvement process (maturity belts) was defined, the engineers started applying both to improve the original requirements specification. After an iterative analysis, the results shown that 66.26 % of the requirements were modified to get a 94 % of high quality requirements in that specification, as shown in Fig. 11. The final specification has 109 new requirements, resulting from splitting many requirements from the original specification that contained more than one need in their statement.

After having improved the specification according to the white belt, we proceeded to analyze the quality results for both specifications according to each of three defined maturity levels. In Fig. 12, it is shown how the quality evolve progressively according to the maturity process, so efforts in the project can be distributed and be more controlled during the whole process.

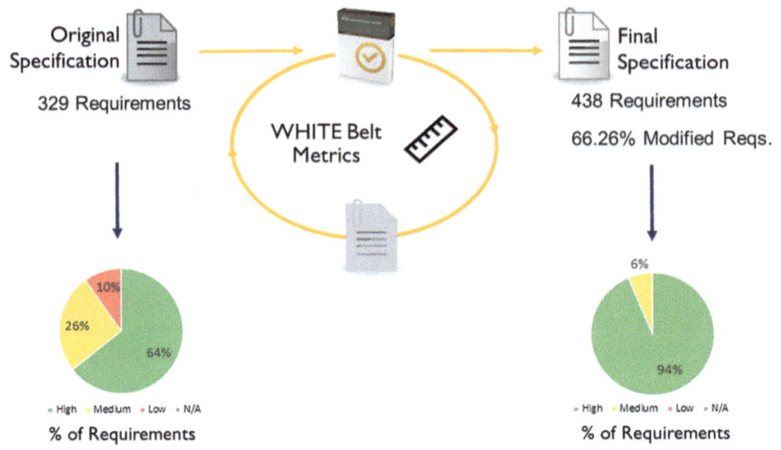

Fig. 11 Quality improvement process results after applying White Belt configuration

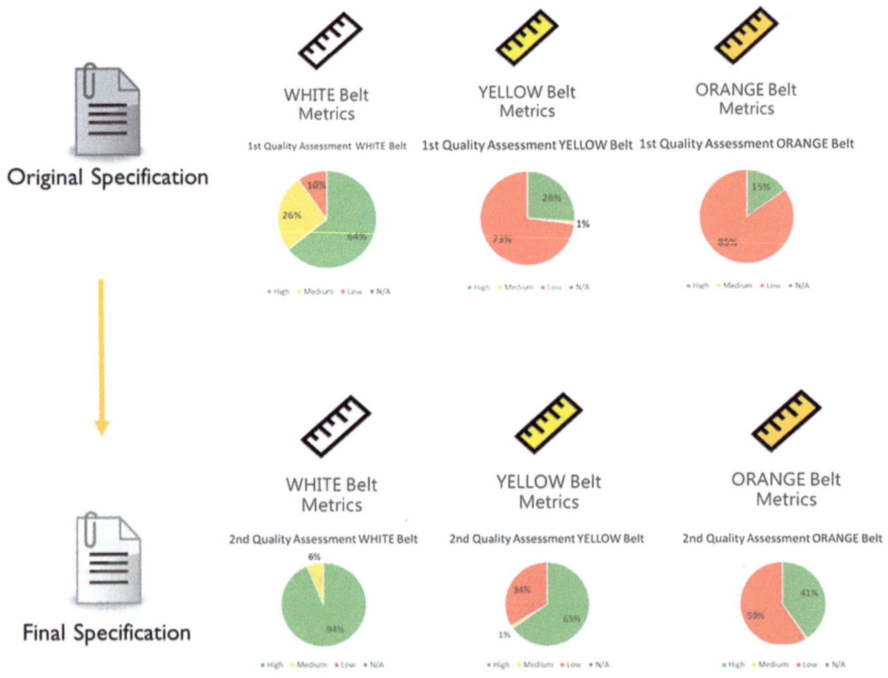

Fig. 12 Quality results for the three different defined levels in the PoC

7 Conclusion

The PoC, as it has been described in this paper seems to be an excellent approach to define and introduce a quality improvement process in a given organization, department or project. Even though the scope of this study is about the railway domain, both the RQS tools and the quality process are suitable for any other domains.

References

1. INCOSE, Systems Engineering Handbook (2015)
2. Martin, J.: An information systems manifesto. Commun. ACM **28**(3) (1985)
3. I. 29148:2011 and ©, "International Standard ISO/IEC/IEEE Systems and software engineering—agile environment," (2011)
4. Génova, G., Llorens, J.: "Requirements Authoring—Fundamentos del proceso de Ingeniería de Requisitos (v3)." (2015)

Systems Engineering Education for East Africa

Solomon Gebreyohannes, Tadilo Endeshaw Bogale, William Edmonson and Lakemariam Yohannes Worku

Abstract The goal of this paper is to explore the need for Systems Engineering (SE) in East African countries and how best to educate engineers in the field of SE. It provides a comprehensive overview on the usefulness of SE education for East African nations and proposes SE curriculum. Presently SE has been given little attention in East African countries. However, these countries are in the beginning of industrialization with many new mega projects and infrastructure expansions that demands SE professionals. This motivates the need to introduce SE education in the region. Systems engineering education demands the development of SE curriculum, which will be multidisciplinary and considers social and psychological factors by taking into account the active participation of the community. Towards this end, the paper demonstrates the necessity of SE education via designing and managing ongoing mega projects. It also proposes SE curriculum by incorporating courses covering both foundations of SE and the practice of SE. Furthermore, it suggests that these courses will be delivered in collaboration with industry and government entities.

S. Gebreyohannes · W. Edmonson (✉) · L.Y. Worku
NC A & T State University, Greensboro, NC 27411, USA
e-mail: wwedmons@ncat.edu

S. Gebreyohannes
e-mail: shgebrey@aggies.ncat.edu

L.Y. Worku
e-mail: lyworku@aggies.ncat.edu

T.E. Bogale
Institut National de la Recherche Scientifique (INRS), Montreal, QC, Canada
e-mail: tadilo.bogale@emt.inrs.ca

G. Fanmuy et al. (eds.), *Complex Systems Design & Management*,
DOI 10.1007/978-3-319-49103-5_15

1 Introduction

The exponential growth of contemporary large scale and complex systems necessitate an interdisciplinary and organized design approach. In general, most complex systems require careful design consideration a priori, which will have substantial influence on the overall system performance. Systems Engineering (SE) aims at addressing the design, integration, testing, and decision support necessary to develop multidisciplinary, cost effective, scalable and robust complex systems [1–3]. Systems Engineering methodologies is being used in the development of different areas including aeronautic, medical, entertainment and infrastructure projects, such as telecommunication, transportation, building and energy management. On the other hand, most current industries require expertise who can understand and integrate different parts of a system. These diverse application and industrial needs motivate the government and academic institutions to introduce SE as a field of study to produce these multidisciplinary professionals [4–6].

Although the field of SE is relatively new as compared to traditional engineering disciplines, recent trends have shown a growth in graduate programs in the United States, Australia, Europe, and many universities across Asia. However, this academic direction has given little (to no) attention to African universities with the exception of South Africa. The goal of this work is to provide a comprehensive overview on the benefits of SE education for African nations. In particular, we emphasize the necessity of SE education in the countries that make up the Eastern region of Sub-Saharan Africa and how best to educate engineers in the field of SE. One of the main components is to understand systems thinking and how it could be used to solve many of the regions issues through the use of SE methodologies, processes, and tools. This includes the knowledge that it requires multiple disciplines along with stakeholders coalescing around a solution. The use of SE goes beyond just solving problems, but can also be used to develop new or enhance existing products and processes. The later could lead to economic sustainability and an economy less dependent on importation of products and knowledge through the development of an indigenous industrial base.

The paper also explores the need for understanding and involving the end user along with their personal interaction and therefore the community must be an integral part of its development. The systems engineer needs to understand and take into account the cultural aspect from a design and usage standpoint. This could ultimately lead to development of a SE curriculum that will go beyond traditional SE programs. An example of the SE process will be motivated and demonstrated via designing, managing, and planning mega projects for Ethiopia, which is one of the East African countries with fastest economic growth in the world and the second most populous nation in Africa. Such projects involve complex systems to be managed wisely based on SE principles.

The rest of the paper is organized as follows. Section 2 introduces SE, Model-Based Systems Engineering (MBSE), and its theoretical foundation and applications. A SE methodology developed by the authors, the Responsive and

Formal Design (RFD), a design process that represents a merger between MBSE and the mathematical foundation of formal methods. It also outlines the merits of SE in East Africa and shows how SE can be used in planning, designing, managing, operations, and retirement of mega-projects. Section 3 lists engineering programs of East African universities and ongoing projects of the countries. Section 4 presents the SE curriculum design. Section 5 concludes the paper.

2 Background and Motivation

Systems Engineering has been given little attention in Africa with the exception of South Africa and therefore poorly represented in the SE community. There is no East African representation in the INCOSE worldwide directory of Systems Engineering and Industrial Engineering Academic Programs [7], again with the exception of South Africa. On the other hand, these countries are engaged in various engineering projects that warrant the use of SE. Examples of the need for SE are the several mega projects being developed (hydro-electric, communication networks) along with the push for Internet of Things that could involve smaller design projects that integrate into the network, e.g. telephone banking [8]. Systems engineering could help to plan, design, analyze, and integrate these systems, and predict project complexity [4, 6]. These necessitate the need for these countries to explore and use SE principles.

Systems Engineering is defined as:

> … an interdisciplinary approach and means to enable the realization of successful systems. It focuses on defining customer needs and required functionality early in the development cycle, documenting requirements, and then proceeding with design synthesis and system validation while considering the complete problem: operations, cost and schedule, performance, training and support, test, manufacturing, and disposal. Systems engineering considers both the business and the technical needs of all customers with the goal of providing a quality product that meets the user needs (p. 7 [9]).

It seeks a safe and balanced design in the face of opposing interests and multiple, sometimes conflicting constraints [10, 11]. Systems engineering is a multidisciplinary field and its main foundations are systems theory, decision theory, probability theory, abstract mathematics, organizational theory, psychology, behavioral economics, and engineering. Systems engineering fosters the engineers' problem solving capability for a sustainable product development in different sectors including health, education and infrastructure (communications, power). It also promotes "systems thinking" philosophy to conceptual and real models [12].

Systems engineering process can also be considered as "the application of scientific and other knowledge to practical task by ordered systems that involve people and organizations, living things, and machine" [13]. In general, one can envision this as conjunction of the following aspects: technological, cultural, and organization.

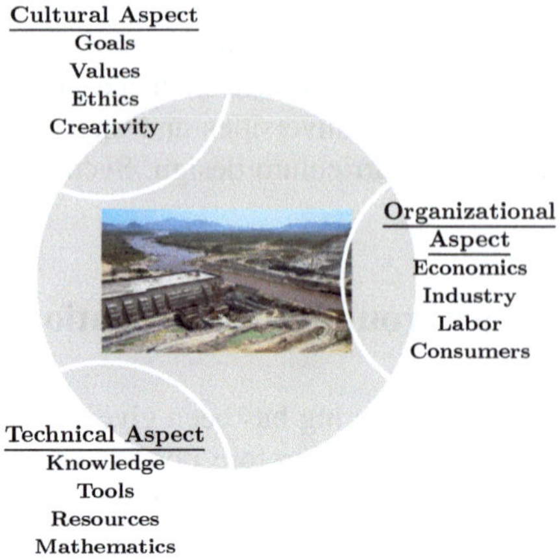

Fig. 1 Technology Process Definition (Center figure of Ethiopian dam [14])

See Fig. 1 for further clarification. Note that it is the cultural aspect as defined by Pacey is normally not addressed by most SE programs or organizations, but provides a unique perspective to how different cultural groups interact with technology.

A recent development in SE aims to represent the system components as models that are an abstracted representation of reality, defined as Model-based Systems Engineering (MBSE). The MBSE methodology is about elevating models in the design process to a central and governing role in the specification, design, integration, validation, and operation of a system. This is a paradigm shift from traditional document-based and acquisition life-cycle model approaches [15]. There is a recent interest in developing a theoretical foundation for MBSE that is based strongly in the field of mathematics. The motivation for this development is to provide a methodology for the design of complex systems which its process is correct with respect to requirements. Correctness is defined with respect to having consistency (no contradiction) at each abstraction layer and to prove traceability with each refinement of the design. Interest amongst the community is growing, as is evident from the attendance of Theoretical Foundations of Systems Engineering Special Sessions at the Annual IEEE International Systems Conference and the International Symposium on Systems Engineering. Several of the authors of this paper have proposed a Responsive and Formal Design (RFD) process that integrates MBSE with formal methods that results in correct design as defined previously [16, 17] which shows a comprehensive use of mathematics [18–20]. A graphical description is shown in Fig. 2, which represents a layout for each level of abstraction that makes up the project definition portion of the SE V model [9, 10].

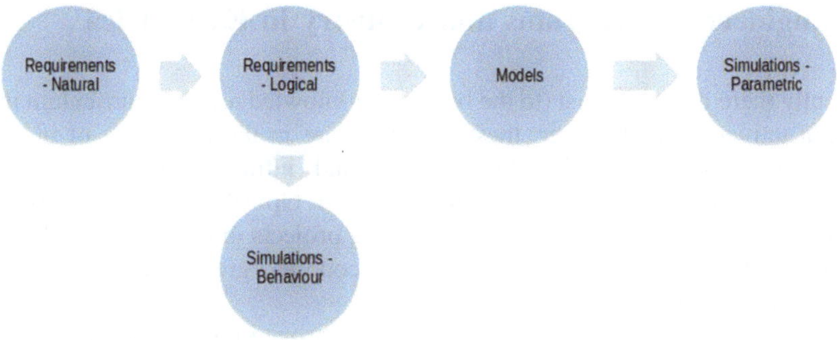

Fig. 2 The responsive and formal design process

Systems engineering has a long history of applying these methodologies to the design of systems, but has a minor presence in East African countries as a tool for their development. Nowadays East African countries are participating in mega-engineering projects, e.g., space program [21], infrastructure development and hydro power projects to secure high growth rates over a longer period of time [22]. These projects require SE in its design and management. However, since there is a lack of SE professionals in the region, efficient planning, designing, and managing are not practiced for these projects, which could lead to economic loss and extended development times; e.g., infrastructures built in the countries can render useless years later due to construction of new projects. Though these projects require SE, it is usually outsourced to foreign companies and therefore indigenous skill set is not developed.

Ethiopia is one of East African countries with fastest economic growth in the world. In recent years, Ethiopia's development has been supported by the expansion of infrastructures such as hydropower dams, roads, electrical installations, telecommunications system, etc. [22–24]. However, there is a huge challenge for different institutions to find a systematic way to work together on planning, designing, operation, and retirement of these systems. One example is the organizations of EthioTelecom, Road Authorities, Water and Sanitation Service, and Electric service agencies often fail to cooperate, whereby in the capital city, the road which costs more than 25 million USD was replaced by light railway for the city after a couple of years of service. Systems engineering could play significant role to prevent these kinds of losses through the involvement of all organizational stakeholders. This in turn contributes to establishing a more sustainable and integrated developments in Ethiopia, as outlined in Fig. 1. It is also important to establish systems thinking and practice as part of the post-secondary education systems and in the development of a graduate SE curriculum.

3 Engineering Programs and Industry in East Africa

Presently there does not exist (to the best of our knowledge) a SE curriculum in East African universities. However, these countries are in the beginning of industrialization with many emerging big projects and infrastructure expansions that demands SE professionals. This section presents typical list of engineering programs in East African universities and ongoing projects in the countries. The focus will be on those countries with significant economic growth rates, i.e. Rwanda, Uganda, Kenya, and Ethiopia.

Table 1 represents a list of some public and private universities in East Africa with engineering programs. This data is taken from the universities website.

Traditionally, SE reside in Departments of Industrial Engineering, but there is usually no cross pollination with other engineering departments even though modern engineering project require multi-discipline solutions. This becomes particularly true as the product or process becomes more complex or the solution space crosses multiple domains, e.g. cyber-physical systems which consist of computing,

Table 1 List of prominent universities in East African countries of Rwanda, Uganda, Kenya, and Ethiopia

	University	Undergraduate Eng. Programs	Graduate Eng. Programs
Rwanda	Univ. of Rwanda	Civil, Water & Env, Electrical, Mechanical, Surveying & Construction Tech, Building & Construction Tech, Electronics, Telecom, and Computer.	Transportation and Economics, Highway Eng. and Mgt.
Uganda	Makerere Univ.	Software, Agricultural, Civil, Electrical, Mechanical, Telecom, Computer, and Biomedical.	Agricultural, Data Comm & Software, Civil, Elect, Mech, Renewable Energy, Power Systems, Telecom, Urban Planning & Design.
	Kampala Int. Univ.	Electrical, Civil, Mechanical, Telecom, Computer, Software.	Electrical, Water & Env, and Software Systems.
Kenya	Univ. of Nairobi	Civil, Electrical, Mechanical, Geospatial, Env & Biosystems, and Electrical and Electronic.	Mechanical, Electrical and Information, Env and Biosystems, & Civil.
	Moi Univ.	Civil & Structural, Chemical & Process, Electrical & Telecom, Mechanical & Production, Industrial & Textile.	Structural, Water, Chemical, Textile, Industrial, Production, Env Eng.
	Jomo Kenyatta Univ. of A&T	Civil, Electrical & Electronic, Electrical & Computer, Soil, Water, & Env, Mechantronics, Mechanical, Telecom & Information, Geomatics, Biomedical & Process.	Agricultural Processing, Bio-Systems Structural, Mechanical, Software, Soil & Water, Telecom, Biomechanical, Processing & Structures, Soil, Water, & Env.

(continued)

Table 1 (continued)

	University	Undergraduate Eng. Programs	Graduate Eng. Programs
Ethiopia	Addis Ababa Univ.	Electronic Comm, Power, Industrial Control, Computer, Microelectronics, Chemical & Bio, Civil & Env, Mechanical.	Industrial, Mechanical Design, Thermal, Railway, Env, Process, Food, Bio-Medical, Material.
	Jimma Univ.	Civil, Chemical, Electrical & Computer, Mechanical, Biomedical, Hydraulic & Water Resources, Water Supply & Env, Material Science.	Sustainable Energy, Structural, Water Resources, Env, Electrical Power, Geotechnical, Construction & Mgt, Hydraulics, Comm, Material Science, Polymer, Metallurgy.
	Bahir Dar Univ.	Textile, Garment, Leather, Chemical, Food Tech & Process, Civil, Water Resources & Env, Mechancial, Industrial, Electrical.	Env, Hydraulic, Geotechnical, Manufacturing, Process, Water Resources, Power Systems, Production & Mgt, Structural, Thermal, Sustainable Energy, Textile Manufacturing.

controls, and/or monitoring of the systems environment. Examples of cyber-physical systems are medical devices, water management, auto pilots, etc.

East African countries are in the beginning of industrialization with many new big projects and infrastructure expansions that demands SE professionals. In other words, how does Africa become part of the 4th Industrial Revolution (Information Age) where technology is the key driver. The growth of information technology over the last several decades and the prospects for continued rapid growth in this field have precipitated the need for systems engineers who can design and integrate large-scale information-systems for enterprises [25]. As stated by President Alpha Conde of Guinea at the World Economic Forum Africa 2016 "Emerging nations … can make better choices about how they develop. We don't need to repeat the mistakes of western countries." This view should be the driver on not only the projects that need to be developed but also determine the design of post-secondary education [26]. Below is a list of several major projects that can greatly benefit from indigenous SE.

- *Space Science and Technology Research and Electric Power Generation and Distribution projects in Ethiopia:* Entoto Observatory and Research Center is a research and training institute in Addis Ababa whose plan is the implementation of space-earth observing systems and related sciences, e.g., construction of ETHIOSAT satellite [21]. This lays the foundation for Ethiopia's further advancement for autonomy in space technology and science. ETHIOSAT function is to photograph the earth with additional remote sensing capability. There are also different electricity generation projects in Ethiopia such as hydropower, wind farm, and geothermal energy projects. Ethiopia is constructing a dam (Grand Ethiopian Renaissance Dam, GERD) that will generate 6,000 MW of power and represents the largest hydroelectric power plant in Africa when completed [14].

- *Mega Projects in Kenya, Uganda and Rwanda:* Consist of: (1) Kenya have been implementing mega projects with a value of $60 billion since 2014. The projects include Standard Gauge Railway (SGR) and different power transmission and distribution stations under construction [27]. (2) Uganda faces trade challenges due to poor infrastructure. To address these challenges, the country is engaged with infrastructure and human resource development projects of railways, roads, oil pipelines, and inland waterway transportation projects [28]. (3) Rwanda has been praised by many political and economic analysts for creating a very dynamic and suitable environment for development and investment. The ongoing projects include hydropower and solar power production and railway project (with Rwanda and Burundi) [29].

4 Systems Engineering Education for East Africa

The growing SE research lead the academic community to push for the introduction of formal programs in SE; for example, a number of universities in the United States have added programs related to SE over a period of three decades [30]. Here system is inclusive of the various system types, such as complex and system of systems; each consisting of multiple engineering and organizational disciplines. This makes the SE education complicated by the broad mandate of modern system architectures, the complexity and interrelationship of many constituents, and the relationships with other disciplines throughout the entire system life cycle. This shows that careful educational framework is needed for the SE field of study [30–36]. The demand for SE education in East Africa will require development of a SE curriculum that will be multidisciplinary and possible culturally specific.

4.1 Competencies

Systems thinking, seeing the whole, is what makes SE different from other engineering disciplines. It is developed through experience, education and training. There are some requirements to assess a systems engineer (called 'competencies' [37, 38]). In [38], some SE competency models are presented with a major discussion around the Capacity for Engineering Systems Thinking (CEST) model. The competencies associated with CEST are: cognitive competencies, skills/abilities, behavioral competencies related to knowledge and experience. The cognitive competencies represent understanding the whole system, seeing the big picture, understanding interconnections, closed loop thinking, understand system synergy (emergent properties), understand the system from multiple perspectives, think creatively, understand systems without getting stuck on details, tolerance for ambiguity and uncertainty, having the ability to "see" the future, etc. We refer the reader to [38] for the full competency list and comparison with the other models. To achieve these competencies, SE education curriculum need to be well organized and designed considering the regions resources, and social and psychological factors.

4.2 Curriculum

There are different factors affecting the design of a SE education curriculum. The learning and teaching processes of SE are dominated by social and psychological factors [6]. On this aspect, East African countries have their own cultural background and custom. Hence, SE education curriculum design may need to consider these factors by taking into account the active participation of the community and their cultural views, but still following the general outline of SE programs in other countries. Systems engineering education need to be delivered as an interdisciplinary academic program in collaboration with industry and government entities [32]. The SE curriculum need to have courses covering both foundations of SE and the practice of SE. Foundations of SE could consist of: mathematics, decision theory, probability theory, organization theory, and engineering courses associated with a particular expertise. While SE design and practice courses could consist of: system architecting, risk management, requirements engineering, trade-space analysis [39]. Since SE study covers a wide range of areas and its skills derive from other branch of engineering, it might be appropriate to establish a graduate level or as a joint degree program with other engineering departments, e.g. electrical, mechanical, etc. It is anticipated that students with engineering background are the target group where they can potentially apply their knowledge to industry and community service sectors. The SE curriculum will have its own mission, focus, objectives, and list of courses and projects whose primary mission is to assist students in achieving full professional competence by offering relevant courses in conjunction with industry exposure through projects and internships. As an example, a brief review on fundamental concepts related to systems theory and SE of electronics is presented in [40]. The SE curriculum also needs to meet the general objectives set by the countries' Ministry of Education to prepare students for career opportunities in industry and public services.

Courses and Projects. The SE curriculum needs to have prerequisite, core, and elective courses, mini and final projects, and industry internship. The curriculum can be broken into the following categories of courses:

- Prerequisite courses may consist of engineering, science, mathematics, psychology, economics, history, and culture courses.
- Core courses may consist of foundations of SE, overview of SE, and architecture of SE, decision theory, probability theory, organizational theory, behavioral economics, engineering, and software tools.
- Industry internship helps the students to apply their knowledge and obtain some experience with SE projects. The students need to do a project supervised by a

mentor (from the industry) and an advisor (a professor from a university). It is preferred for the students to work within a group setting [41].
- Elective courses will cover specific topics and software tools that is important to their thesis.
- Engineering departments will require that the students to do a master's thesis for graduation.

5 Conclusion

Systems engineering is an emerging and interdisciplinary field, which is used to design and manage complex engineering systems. Hence, there is a need of SE education for engineers in general. However, little attention is given in African universities. In this paper, we outline the necessity of SE education for East Africa in particular and show how to educate engineers in the region through a discussion of development of a SE curriculum.

References

1. Elm, J.P.: Developing a business case for systems engineering. Aerosp. Electron. Syst. Mag. IEEE **27**(7), 13–19 (2012)
2. Ramos, A.L., Ferreira, J.V., Barcelo, J.: LITHE: an agile methodology for human-centric model-based systems engineering. IEEE Trans. Sys. Man. Cyber. **43**(3), 504–521 (2013)
3. Rao, B.H., Padmaja, K., Gurulingam, P.: A brief view of model based systems engineering methodologies. J. Eng. Trends Technol. (IJETT) **4**(8), 3266–3271 (2013)
4. Sage, A.P.: Systems engineering education. IEEE Tran. Syst. Man. Cyber. **30**(2), 164–174 (2000)
5. Seymour, S.J., Luman, R.R.: Academic perspectives of systems engineering. John Hopkins Tech. Digest. **29**(4), 377–386 (2011)
6. Asbjornsen, O.A., Hamann, R.J.: Toward a unified systems engineering education. IEEE Trans. Sys. Man. Cyber. **30**(2), 175–182 (2000)
7. Systems Engineering Research Center: INCOSE Worldwide Directory of Systems Engineering and Industrial Engineering Academic Programs (2016)
8. The Economist: Why does Kenya lead the world in mobile money? [Online] http://www.economist.com/blogs/economist-explains/2013/05/economist-explains-18
9. INCOSE SE Handbook Working Group: INCOSE Systems Engineering Handbook, v. 3.2. 2. Technical report, INCOSE-TP-2003-002-03.2. 2 (2011)
10. NASA Systems Engineering Handbook, NASA/sp-2007-6105 (2007)
11. ISO/IEC 15288 International Standard on Systems and Software Engineering—System life cycle processes (2008)
12. Rogers, E.M.: Diffusion of innovations. Simon and Schuster (2010)
13. Pacey: The Culture of Technology. MIT Press (1984)
14. Salini Impregilo: Grand Ethiopian Renaissance Dam Project. [Online] http://www.salini-impregilo.com/en/projects/in-progress/dams-hydroelectric-plants-hydraulic-works/grand-ethiopian-renaissance-dam-project.html

15. Survey of Model-Based Systems Engineering (MBSE) methodologies. INCOSE-TD2007-003-02 (2008)
16. Edmonson, W., Herencia-Zapana, H., Neogi, N., Moore, W., Ferguson, S.: Highly confident reduced life-cycle design process for small satellite systems: Methodology and theory. In: Complex Systems and Data Management Conference (CSDM) (2012)
17. Edmonson, W., Chenou, J., Neogi, N., Herencia-Zapana, H.: Small satellite systems design methodology: a formal and agile design process. In: 8th Annual IEEE International Systems Conference, pp. 518–524 (2014)
18. Gebreyohannes, S., Edmonson, W., Chenou, J., Neogi, N., Esterline, A.: Formal requirement management for the responsive and formal design process. In: First IEEE International Symposium on Systems Engineering, pp. 364–369 (2015)
19. Chenou, J., Edmonson, W., Esterline, A., Neogi, N.: Formal framework for ensuring consistent system and component theories in the design of small satellite systems. In: Complex Systems and Data Management Conference (CSDM) (2014)
20. Gebreyohannes, S., Edmonson, W., Esterline, A., Chenou, J.: Requirement hierarchy for the responsive and formal design process. In: Second IEEE International Symposium on Systems Engineering (2016)
21. Ethiopian space science society (ESSS). [Online] http://www.ethiosss.org.et/index.php/en/
22. Foster, V., Morella, E.: Ethiopia's infrastructure: a continental perspective. In: World Bank Policy Research Working Paper Series (2011)
23. East Africa: Ethiopia's huge infrastructure projects whet continental banks' appetite. [Online] http://allafrica.com/stories/201511090663.html
24. International Growth Center (IGC): Urban experimentation: how housing, transport, and infrastructure projects are revolutionizing Addis Ababa. [Online] http://www.theigc.org/blog/urban-experimentation-how-housing-transport-and-infrastructure-projects-arerevolutionising-addis-ababa/
25. Brown, D.E.: Information technology as a component of system engineering education. In: IEEE International Conference on Systems, Man, and Cybernetics, vol. 4, pp. 3132–3137 (1996)
26. World Economic Forum: How the technology revolution will transform Africa? [Online] https://www.weforum.org/agenda/2016/01/what-does-the-fourth-industrial-revolution-mean-for-africa/
27. Mega Projects in Kenya. [Online] https://www.megaprojects.co.ke/
28. Trade Mark East Africa. [Online] https://www.trademarkea.com/countries/uganda/
29. Africa Details: Multi-Billion Railway Project in Rwanda. [Online] http://www.africadetails.com/index.php/news/80-building-interiors/670-multi-billion-railway-project-in-rwanda
30. Brown, D.E., Scherer, W.T.: A comparison of systems engineering programs in the United States. IEEE Trans. Syst. Man Cybern. Part C: Appl. Rev. **30**(2), 204–212 (2000)
31. Faulconbridge, I., Ryan, M.: A framework for systems engineering education. In: Systems Engineering Test & Evaluation Conference, pp. 3–7 (1999)
32. Arias, J.M.: A Systems Engineering Educational Framework for Economic Growth in Spain within a European Centered Aerospace Market
33. Subramanian, T., Dubey, P. et al.: Systems engineering: a new approach to engineering education in India. In: IEEE International Conference on Engineering Education: Innovative Practices and Future Trends (AICERA) (2012)
34. Keys, L.K.: Systems engineering and technology management education for the 21st century. In: Portland International Conference on Management of Engineering and Technology (PICMET), IEEE, pp. 2152–2170 (2009)
35. Hilburn, T.B., Squires, A., Madachy, R.: A model for educating systems engineers. IEEE International Systems Conference (SysCon), pp. 1–5 (2012)
36. Troutman Jr., B.L.: A company-sponsored program for systems engineering education. Proc. IEEE **66**(8), 969–972 (1978)

37. Frank, M.: Assessing the interest for systems engineering positions and other engineering positions' required Capacity for Engineering Systems Thinking (CEST). Syst. Eng. **13**(2), 161–174 (2010)
38. Frank, M., Kasser, J.: Assessing the Capacity for Engineering Systems Thinking (CEST) and Other Competencies of Systems Engineers. INTECH Open Access Publisher (2012)
39. Chris, P.: NSF Program Overview: Engineering & System Design (ESD) and Systems Science (SYS). 1.13 (2014). [Online] http://www.nsf.gov/eng/cmmi/documents/NSF_ProgramBriefing_v1.13_20141202.pdf
40. Castro, M., et al.: A system theory perspective of electronics in engineering education. In: Education Engineering (EDUCON), pp. 1829–1834 (2010)
41. Shimazu, K., Ohkami, Y.: Systems engineering education for inexperienced students by providing hand-on practices. In: IEEE International Systems Conference (SysConf) (2011)

Systems Engineering Human Capital Development: Objectives and Research Directions

Jon Wade

Abstract This paper presents the challenges in ensuring the existence of a workforce this is capable of conceiving, realizing and supporting increasingly complex systems throughout the lifecycle. Accomplishing this will require that all systems decision makers are systems thinkers, all engineers have systems engineering skills, and all systems engineers are broad-based technical leaders. The human capital development and academic forum research focus areas of the Systems Engineering Research Center (SERC) and the International Council on Systems Engineers (INCOSE), respectively, are presented in this work. In addition, current status and future efforts are discussed.

1 Introduction

Over the last decade, the US Department of Defense, defense industrial base and commercial industries have often cited a shortfall in the quantity of systems engineers and in the knowledge, skills, and abilities of those systems engineers. Not only is there a critical shortage of systems engineers, but the skill sets and capabilities of these engineers need to rapidly expand to address the growing complexity in the systems they are attempting to engineer. Systems Engineering Vision 2025 presents a future view of SE [1]. That report highlights several areas that directly impact Human Capital Development.

Systems Engineering is not only for those with the title of "Systems Engineer". Systems skills are essential for systems decision makers, technical leaders and all engineers. All leaders and those making decisions about systems need to be systems thinkers. Systems thinking skills need to be developed long before graduate studies and should be introduced as early as kindergarten through high school. All engineers should have some education and training in systems and SE. While under-

J. Wade (✉)
Stevens Institute of Technology, Castle Point on Hudson,
Hoboken, NJ 07030, USA
e-mail: jon.wade@stevens.edu

G. Fanmuy et al. (eds.), *Complex Systems Design & Management*,
DOI 10.1007/978-3-319-49103-5_16

graduate curricula are already full, these skills can be introduced and distilled in cornerstone and capstone projects. Finally, systems engineers need to be well versed in a broad set of socio-technical and leadership skills, serving as a central, multi-disciplinary focal point of systems development with stakeholders of all types.

While those with the title of *Systems Engineer* may hold numerous roles, e.g., 17 roles are enumerated in the Helix research [2], there is a smaller set of roles that can be attributed to the specific work of systems engineering and the skills that it entails. These roles have been divided into those that are focused on the system being developed, and those that are focused on the system that is doing the developing. Starting with the existing set of 17 roles, the following is a set that have been identified to be related to the act of Systems Engineering, as shown in Table 1.

The roles of Concept Creator and Support Engineer, the front and the back-end of the lifecycle, were missing from the original list of 17 roles. (Eliminated from the list of 17, were the roles of Functional, Technical and Information Managers, Coordinators, Instructors and the ambiguous 'Classified Ad'.)

This notion of systems engineering both participating in and managing the life-cycle provides some clarity of the activities of systems engineering. Whereas Systems Engineers are often seen as "process people", in fact systems engineering involves both process and the actual activities enabling systems to progress through the lifecycle. This is consistent with the view that all engineers need to have systems engineering skills. While all engineers might not manage the processes or systems responsible for the lifecycle, they certainly will be participating in the lifecycle itself.

The four step framework of Conception, Design, Implementation and Operation (sustainment) is the basis of the worldwide CDIO movement. "CDIO is based on a commonly shared premise that engineering graduates should be able to: Conceive–Design–Implement–Operate complex value-added engineering systems in a modern team-based engineering environment to create systems and products" [3].

Table 1 Systems engineering roles

FOCUS: system being developed	FOCUS: process and organization
Concept development	*Process development*
Concept creator	Process engineer
Requirements owner (create and maintain)	Logistics/Operations
Systems architecture and design	*Management/Communications*
System architect ("system designer" + "glue")	Program/Project management
Systems analyst	Stakeholder interface
Implementation	
Detailed designer	
Verification and Validation engineer	
Support and sustainment	
Support engineer	

It is the vision of CDIO of an education that stresses the fundamentals with the following properties:

- A curriculum organized around mutually supporting courses, but with CDIO activities highly interwoven
- Rich with student design-build-test projects
- Integrating learning of professional skills such as teamwork and communication
- Featuring active and experiential learning
- Constantly improved through quality assurance process with higher aims than accreditation

The awareness of the interconnections between decisions made throughout the lifecycle are an integral part of the maturation of engineering developing systems and systems engineers responsible for overall system success. As a result, *experience acceleration*, as described in the SERC research strategy, involves providing students and practitioners with experience in which they are made aware of these relationships. This can be accomplished through experiential learning in the form of simulations or hands-on stepping stone and capstone projects.

In addition, systems engineering is a discipline that can be applied across domains and system types. However, it may be difficult for many to make what is likely to be a very abstract translation between these domains. The Cynefin framework, shown in Fig. 1, is illustrative of the very difficult systems engineering approaches that are required for simple, complicated, complex and chaotic systems.

This paper first describes the research objectives, strategies and research program for the Systems Engineering Research Center (SERC) University Affiliated Research Center (UARC) Human Capital Development (HCD) research area that directly targets the aforementioned shortfalls and challenges [4]. It then describes the research approach and questions raised by the INCOSE Academic Forum. Finally, it concludes with thoughts on the future of research in the education and training of Systems Thinking and Engineering.

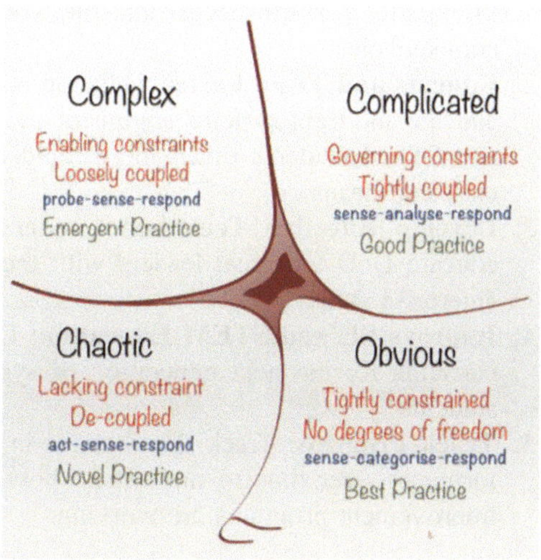

Fig. 1 Cynefin framework

2 SERC Strategies

The Systems Engineering Research Center (SERC), a University-Affiliated Research Center of the US Department of Defense, leverages the research and expertise of faculty, staff, and student researchers from more than 20 collaborating universities throughout the United States. Led by Stevens Institute of Technology and principal collaborator, University of Southern California (USC), the SERC has engaged more than 400 researchers since its founding in 2008—a community of broad experience, deep knowledge and diverse interests. SERC researchers have worked across many domains and industries, including finance, telecommunications, computing, transportation, in addition to defense, enabling them to bring broad perspectives to their research.

This Systems Engineering Research Center (SERC) 2014–2018 Technical Plan describes the SERC Vision, the Sponsor's needs, and the SERC's response to these needs, supported by research in Enterprises and Systems of Systems (ESOS), Trusted Systems (TS), Systems Engineering and Systems Management Transformation (SEMT) and Human Capital Development (HCD). The following is the goal for the HCD area:

Ensure a competitive advantage through the availability of highly capable systems engineers and technical leaders for the DoD and the defense industrial base.

The HCD Grand Challenge to achieve the HCD goal is to:

Discover how to dramatically accelerate the professional development of highly capable systems engineers and technical leaders in DoD and the defense industrial base and determine how to sustainably implement those discoveries.

It is believed that successfully executing the following strategies will make significant progress towards addressing the HCD Grand Challenge:

1. **Create and Provide Easy Knowledge Access**: Make it easy for systems engineers to understand the SE discipline and to access the information needed to expertly perform SE so that the workforce can master the most important competencies
2. **Educate and Train Faster**: Develop innovative approaches and technology to educate and train systems engineers and systems teams at all levels, engineers, and STEM students much more rapidly, effectively and efficiently than with classical means
3. **Develop Effective Technical Leaders**: Develop innovative approaches to educate DoD technical leaders with the right mix of technical, business, and enterprise skills
4. **Improve SE and STEM Education**: Develop recommendations and systems curricula for the next generation of systems engineers, engineers and STEM students
5. **Track Progress**: Track the changes in SE workforce demographics and performance over time to understand how the workforce is improving and how improvement programs are working

Three HCD research programs directly implement the strategy:

- Evolving Body of Knowledge
- Experience Acceleration
- Systems Engineering and Technical Leadership Education

These research programs and their constituent projects are described in the next section.

3 SERC Research Programs

3.1 Evolving Body of Knowledge Program

This research program primarily implements HCD strategies 1, 4 and 5 above—Create and Provide Easy Knowledge Access, Improve SE and STEM Education, and Track Progress. It includes two current projects—Helix and SEEK. A third project—BKCASE—was successfully completed at the end of 2013 as a research effort, although the SERC maintains a role as one of three stewards leading the operation and maintenance of BKCASE products.

BKCASE
The Body of Knowledge and Curriculum to Advance Systems Engineering (BKCASE) Project is (a) identifying and making readily accessible the vast knowledge that systems engineers need to know (SEBoK) and (b) providing recommendations to the SE academic community on SE graduate curricula (GRCSE). BKCASE began in 2009 as a SERC-supported project led by Stevens Institute and the Naval Postgraduate School. Beginning in 2013, INCOSE and the Institute of Electrical and Electronics Engineers Computer Society (IEEE-CS) became co-stewards with the SERC to guide and promulgate the SEBoK and GRCSE. Both products undergo regular updates to reflect advances in the field and feedback from the user community. SEBoK articles have been accessed more than 1,000,000 times since Version 1.0 was released in September 2012. Several universities in the US, Europe, and elsewhere have been adopting GRCSE curriculum recommendations. The SEBoK is novel in its form of delivery (a wiki), its governance model (shared among three organizations), its scale (spanning the technical aspects of the discipline, how that technology is effectively adopted and used, and the underlying science on which the technology is based), and its rate of change (multiple updates annually).

Helix
Helix began in October 2012 to examine the "DNA" of the systems engineering workforce in both DoD and the defense industrial base. The project is addressing three research questions:

- What are the characteristics of systems engineers?
- What enables them to be effective and why?
- What are employers doing to improve the effectiveness of their systems engineers?

Based on interviews with nearly 300 systems engineers and those who work with systems engineers, Helix developed Atlas, a theory of what enables systems engineers to be effective. Atlas describes the key proficiencies that impact the effectiveness of systems engineers, the several forces that impact the level of proficiency that systems engineers obtain, how the career paths of systems engineers progress, how personal and organizational characteristics affect the evolution of systems engineers, and also provides demographic data about systems engineers, such as their typical education and how that demographic data has changed over time. The project has expanded beyond systems engineers in the defense community and is now looking more broadly at systems engineers in such commercial sectors as healthcare and information technology.

SEEK (Systems Engineering Expert Knowledge)
This project addresses a gap in the SE research literature: the lack of detailed case studies about SE successes and failures. The research developed its first case studies in 2014 and 2015, tailored to defense education needs. These case studies will support instruction at DAU and at the Naval Postgraduate School, the federal service academies, and other government education and training providers.

Much of SE and technical role documentation provides a description of the "who", "what" and "when". However, there is very little guidance on the "how" for these activities. Case studies are a tool that can be used to provide realism and bring the systems engineering practice to the classroom. They can be a valuable source of lessons learned and underscore the effect of decision making. The intent of this work is to provide the means to support the teaching of the "how" through the use of case studies and program artifacts. Rather than adapting a case study to a course curriculum, these cases have been selected and developed with the end result in mind, namely supporting the new DAU Reliability & Maintenance course. The new course consists of five topics covering a significant (e.g. Acquisition Category 1 and 2) Defense Acquisition program lifecycle. The approach is to create a set of two companion case studies that cover the entire program lifecycle, particularly through Test and Evaluation. To provide contrast, one of the case studies will describe a relatively successful program, while the other will cover one that was not as successful.

3.2 Experience Acceleration Program

This research program primarily implements HCD Strategy 2—Educate and Train Faster. It will include projects aimed at creating automated learning environments that simulate real world experiences of systems engineers. Those experiences will be vivid and realistic enough to significantly accelerate the learning and maturation of those systems engineers. One project will evolve the current simulation platform, making it ever more robust and capable and enabling quicker and easier construction of new experiences. Other projects will add to the current catalog of experiences, developing new experiences that use the simulation platform. Experiences will vary based on the size and types of systems being acquired, the acquisition lifecycle, the novelty of the technology being acquired, and other parameters of interest. Over the five-year period from 2014–2018, other organizations will join the SERC in improving the experience platform and in developing additional experiences, creating a marketplace for experience acceleration.

3.3 Systems Engineering and Technical Leadership Education Program

This research program primarily implements HCD strategies 2 and 3—Educate and Train Faster and Develop Effective Technical Leaders. It currently includes two primary projects: the Engineering Capstone Marketplace Project and the Technical Leadership Project. The Capstone Project has mixed Core and US Special Operations Command funding. Even more extreme than the Experience Acceleration Program just described, all of the investment for the Technical Leadership Project has come from DAU—none has come from Core funding. Nevertheless, it is included here because it has been so critical to the HCD research effort.

Capstone Marketplace
The Engineering Capstone Marketplace (ECM) Project is the evolution of research begun in 2010, which showed that a multidisciplinary senior capstone project could enhance development of systems SE competencies and increase interest in SE. ECM is now in its third year of matching engineering students working on their capstone or senior design with sponsors who provide challenging real world problems and dedicated mentors or subject matter experts. This matching is done through the ECM website. Current efforts include development of the process and infrastructure to affordably scale this approach nationwide and improve how thousands of students are taught engineering across the US.

Technical Leadership Project
The Technical Leadership Project began in 2010 to evaluate the hypothesis that the technical leadership capabilities of high potential, senior DoD systems engineers and technologists could be accelerated through an educational program in technical leadership. This research has resulted in the creation of an innovative approach to educating technical leaders through three lenses: systems, business, and enterprise. A series of three five-day courses have been prototyped, piloted and are in the process of being transitioned to the DAU. Each course contains a series of independent readings, lectures, case studies, and student in-class exercises to accelerate

systems technical leadership learning. The courses take the student from (a) leading systems development in the face of uncertainty and ambiguity to (b) understanding how commercial businesses or organizations accountable for multi-system and multi-customers strategize, operate and measure performance to (c) the technical leadership expectations of an enterprise senior technical leader responsible for assessing and adapting multi-nodal structural and activity-based processes within DoD or commercial enterprises.

Ongoing research supports the transition to the DAU of the three SYS 350 series Systems Engineering Technical Leadership prototype courses to address the need for technical leadership education in parallel with functional training.

3.4 Progress Towards Goals

Significant HCD progress has been made through a mix of Core-funded and non-Core funded projects. The have been a number of successes in the Evolving Body of Knowledge Program. The BKCASE Project has made great strides in organizing information and making it globally accessible and available. This project was successfully completed and transitioned just as this Technical Plan was approved. Since September 2012, there have been nearly 1,000,000 visits to articles on the Systems Engineering Body of Knowledge (SEBoK) wiki and many universities have adopted all or part of the recommendations found in the Graduate Reference Curriculum on Systems Engineering (GRCSE). Their continued use and evolution will provide an up-to-date source for systems knowledge. The Helix project is showing success in understanding what enables systems engineers to be effective, how systems engineers mature, and in characterizing the systems engineering workforce. Several organizations have begun using Helix for their workforce improvement efforts.

The Experience Acceleration (EA) Program has continued to mature and now has a variety of capabilities that should support experiences in numerous domains and in several different single and multi-player modes. There is a great potential for this technology to advance the strategic objective of educating and training faster. Limited pilots have been conducted that both show the potential of the technology and have served to provide feedback in its subsequent development. In addition, a set of prototype tools have been developed that show the potential for tailoring existing experiences and developing new ones. Critical work moving forward is in learning evaluation and the validation of the hypothesis that technology can be used to accelerate learning for systems thinking and engineering. This can be facilitated through the use of the EA in Collaborator university courses and training. In addition, it will be necessary to show that experiences can be efficiently created and modified by the non-research community. Finally, a sustaining open source community is needed to ensure that Experience Acceleration experiences and technology can be supported for widespread deployment.

The Systems Engineering and Technical Leadership Education Program continues to make strides improving technical leadership and SE education, primarily with non-Core funds. The Engineering Capstone Marketplace Project (which is funded by a mix of Core and non-Core funds) is the evolution of research begun in 2010, which showed that a multidisciplinary senior capstone project could enhance development of SE competencies and increase interest in SE. The challenge is in scaling this approach nationwide, to have impact on how thousands of students are taught engineering across the US. The Technical Leadership Project also began in 2010 to evaluate the hypothesis that the technical leadership capabilities of high potential, senior DoD systems engineers and technologists could be accelerated through an educational program in technical leadership. This initial research has spawned several efforts for DAU and the Army. The former research resulted in the creation of an innovative approach to educating technical leaders through three lenses: systems, business, and enterprise. That approach was captured in courses have been prototyped, piloted and are being transitioned to DAU. Again, the challenge is in expanding the offering of these courses to broaden their impact.

There are a number of additional remaining gaps that will be necessary to address the HCD Grand Challenge. Some of these include: how to better capture the knowledge of systems engineers who are nearing or in retirement, how to more closely couple research results to their dissemination in education, and how to expand systems education into kindergarten through high school.

4 INCOSE Educational Research

The research directions in the INCOSE Academic Forum have been driven by the educational focus described in the Systems Engineering Vision 2025 document. In particular, there is the threefold focus on ensuring that all systems decision makers are systems thinkers, that all engineers have systems engineering skills and that all systems engineers are broad-based technical and socio-technical leaders.

4.1 Systems Thinking for All Systems Decision Makers

Ensuring that all systems decision makers have systems thinking skills will involve bringing systems thinking education into early education, e.g., primary and secondary school (in the US known as K-12). While it may not be possible to involve systems thinking directly with engineering in the earliest of grades, it is feasible to use systems thinking to understand social interactions and behaviors. Important research questions include:

- How do we define and measure Systems Thinking?
- How can it be taught at the primary, secondary, undergraduate, graduate and on the job?
- Are systems thinking skills innate? Can this be measured?

4.2 Systems Engineering Skills for all Engineers

The INCOSE Academic Forum has taken a great interest in integrating systems education into undergraduate engineering education in the forms of a continuing set of workshops [5]. In 2015, engineering and systems engineering faculty met at an International Council on Systems Engineering (INCOSE) Academic Forum in May and at the American Society of Engineering Education (ASEE) annual conference in June to discuss the need for integrating systems engineering knowledge into the education of all engineers.

There are a number of important research questions that need to be addressed, including:

- What are the major dominant contexts for Systems Engineering?
- How do we translate critical SE principles to each of these contexts?
- How is teamwork and collaboration taught most effectively?
- How do we best impart SE experience building throughout the lifecycle for all Engineers?
- How can we make this happen in UG education?
- How do we know if we are being successful?

4.3 Systems Engineers as Technical Leaders and Communicators

The value of a systems engineer is often related to his/her ability to make expert decisions. Those who can make expert decisions across a broader range of disciplines and across a longer range of the lifecycle have greater value to an organization. As systems engineers should be able to span many disciplines and oversee the entire lifecycle, they should provide a tremendous amount of value. Those at the top of the technical decision making tree are the "Chief Systems Engineers" who represent the pinnacle of systems engineering skill. Clearly, attaining this level of skill represents a significant achievement obtained through numerous experiences.

Clearly engineering, in general, is a team sport in which successful collaboration is a critical element of success. Thus, the systems engineer needs to be the master communicator and leader, often from a position of influence rather than authority. Given the vast number of skills that systems engineers need to acquire, how do they also become expert in the soft skills as well?

Resultant research questions include:

- What does it take to be a technical leader? How do we measure these skills?
- How do we develop technical leaders?
- What is the right mix of education, training and mentorship?
- How do we know if we are being successful?

5 Conclusion

This paper presented the challenges in ensuring the existence of a workforce this is capable of conceiving, realizing and supporting increasingly complex systems throughout the lifecycle. While significant progress has been made on the SERC strategy, there is still much that needs to be done to achieve our objective of ensuring that all systems decision makers are systems thinkers, all engineers have systems engineering skills, and all systems engineers are broad-based technical leaders.

References

1. Friedenthal, S., Beihoff, B., Nichols, D., Oster, C., Paredis, C., Stoewer, H, Wade, J.: A World in Motion: Systems Engineering Vision 2025, International Council on Systems Engineering (2014). http://www.incose.org/docs/default-source/aboutse/se-vision-2025.pdf?sfvrsn=4
2. Pyster, A., Henry, D., Hutchison, N., Jauregui, C., Clifford, M.: Atlas: The Theory of Effective Systems Engineers, version 0.5. Stevens Institute of Technology. SERC-2015-TR-108, Hoboken, NJ, 1 Dec 2015 (2015). http://www.sercuarc.org/wp-content/uploads/2014/05/Helix-Report-Atlas-0.5-December-2015.pdf
3. http://www.cdio.org/cdio-vision, 15 Jan 2016
4. SERC 2014–2018 Technical Plan: 2016 Update. Stevens Institute of Technology, Hoboken, NJ, 18 Feb 2016. http://www.sercuarc.org/wp-content/uploads/2014/05/SERC-2014-2018-Technical-Plan-FINAL-unlimited-distribution.pdf
5. Squires, A., Looft, F., Virani, S.S.: Integrating Systems Education into Undergraduate Engineering Education plus INCOSE Spring Academic Forum Summary Workshop Report, Report of Workshop delivered at the 2015 American Society for Engineering Education (ASEE) Annual Conference and Exposition, Seattle, Washington, 14 Jun 2015 (2015)

Part II
Posters

System Engineering Education for Confirmed Engineers: The FAIS Case Study—A 6 Years Feedback

Omar Hammami

Abstract System engineering education is major challenge with regard to the current structure and organization of the higher education system oriented towards specialization. Although several programs have emerged in the recent years in France in universities and school of engineers the field remains broadly unknown. The number of young graduates with system engineering education remains low. At the same time, system engineering is increasingly used in industry now beyond the usual defense and aeronautical applications. As a result continuous education and training needs for confirmed engineers have surged in the recent years. In this paper we describe a feedback on the FAIS program for confirmed engineers with an analysis of the various parameters which have clearly contributed to its success. Several trends have emerged as a result of this program. System engineering is in need of theoretical foundations and this is increasingly expressed by system engineering participants. Requirements engineering and architecture were initially major themes of the training with the use of associated softwares (DOORS, IBM System Architect, MEGA). Significant experience have been achieved in coaching trainees in understanding and mastering both the concepts and practice on significant case studies. However, the trend have increasingly be in expectations from trainees for architecture evaluation and mastering architecture complexity. The FAIS program have strongly evolved during the past 6 years and have better adapted to the growing needs of attendees. A turning point have been the adoption of coaching and supervision of attendees on their own case studies. Departing from a common academic style case study to the current projects of DGA attendees have both enriched the training and increased the involvement of attendees in their training. A trusted environment of learning and exchange have also allowed deepening of the system engineering issues raised by the case studies. This have been allowed by the common work origin of the participants. MBSE, architecture modelling and evaluation have increasingly dominated the training and the trend is confirmed in 2016. Continuous improvements have been brought with growing training hours for some topics and diminishing training hours for others. FAIS have

O. Hammami (✉)
ENSTA ParisTech, 828 Bvd des Maréchaux, 91762 Palaiseau Cedex, France
e-mail: hammami@ensta.fr

G. Fanmuy et al. (eds.), *Complex Systems Design & Management*,
DOI 10.1007/978-3-319-49103-5_17

operated as a sliding window on education topics. Multidisciplinary system simulation, mathematical modelling and optimization coupled with simulation are the emerging trends of the program.

Acknowledgments This author is grateful to all those who contributed to the FAIS training, partners from companies, attendees for the rich exchanges and discussions. FAIS is as good as its ecosystem. The author would like to thanks DGA for their continuous support and for the exchanges. We are grateful to MEGA company for their support with HOPEX tool and IBM for the Academic initiative program.

Integration of Systems Engineering Approach in Product-Lifecycle-Management by Means of a Mechatronic System

Vahid Salehi, Lukas Burseg, Kristin Paetzold, Abdo Chahin, Jihad Taha and Thomas Rieger

Abstract To achieve the full potential of PLM in Systems Engineering tools especially in view of the system's complexity in industries such as the consumer industry a clear understanding of how best to use such systems is important to product development activities. Systems Engineering is an interdisciplinary field of engineering that focuses on how to design and manage complex engineering systems over their life cycles. Issues such as reliability, logistics, coordination of different teams (requirements management), evaluation measurements, and other disciplines become more difficult when dealing with large or complex projects. Systems Engineering deals with work-processes, optimization methods and tools in such projects. It overlaps technical and human-centered disciplines such as control engineering, industrial engineering, organizational studies, and project management. Systems Engineering ensures that all likely aspects of a project or system are considered, and integrated into a whole. After a short introduction, this paper, which is based on the results of the accomplished descriptive study and literature survey of the Design Research Methodology according to Blessing and Chakrabarti, presents a generic integrated approach of System Driven Product Development (SDPD) and demonstrates the general requirements of a generic integrated approach during the Engineering Design of Systems. The second section presents a new approach of Systems Engineering, which is based on SDPD and will explain the different phases and sub-phases of the developed approach. By means of designing an electric skateboard the different phases of the developed generic

V. Salehi · L. Burseg (✉) · J. Taha · T. Rieger
Munich University of Applied Sciences, Lothstr. 34, 80335 Munich, Germany
e-mail: burseg@hm.edu

V. Salehi
e-mail: salehi-d@hm.edu

K. Paetzold · A. Chahin
Universität der Bundeswehr München, Werner-Heisenberg-Weg 39, 85577 Neubiberg, Germany
e-mail: kristin.paetzold@unibw.de

A. Chahin
e-mail: abdo.chahin@unibw.de

G. Fanmuy et al. (eds.), *Complex Systems Design & Management*,
DOI 10.1007/978-3-319-49103-5_18

integrated approach will be demonstrated and presented. Section three will discuss the results of the Prescriptive Study and address the most important issues. In general, this paper presents the Prescriptive Phase of the Design Research Methodology.

Performance Analysis of SDL Systems

Mihal Brumbulli and Emmanuel Gaudin

Abstract The increasing complexity of software systems is constantly fueling the interest in pragmatic analysis methods. These are by no means scarce, but their applicability requires additional expertise that often has a weak relation with the development process or the domain the system is intended for. The model-driven paradigm addresses this issue at a certain extent by raising the level of abstraction closer to the domain and facilitating development and analysis by means of automation. It tries to shift the inherent complexity from the model towards the automation process. Although this has proven to be quite effective in handling functional aspects, the same cannot be stated with confidence regarding non-functional aspects like performance. In this paper we present a model-driven approach for performance analysis based on standardized languages. The functional aspects of the system are captured using SDL and enriched with performance annotations. Available resources are assigned to system components via deployment diagrams, and real test cases described in TTCN-3 drive model execution. Different scenarios can be executed automatically, and the graphical presentation of results can aid the user to decide on the best allocation of resources in terms of execution time and payload.

M. Brumbulli (✉) · E. Gaudin (✉)
PragmaDev, 18 Rue Des Tournelles, 75011 Paris, France
e-mail: mihal.brumbulli@pragmadev.com

E. Gaudin
e-mail: emmanuel.gaudin@pragmadev.com

G. Fanmuy et al. (eds.), *Complex Systems Design & Management*,
DOI 10.1007/978-3-319-49103-5_19

Prerequisites for the Modelling and Analysis of a Product Development Process Using Network Theory

Abdo Chahin, Julian Hoffmeister, Kristin Paetzold and Vahid Salehi

Abstract Network models have already been used with the intent to gain additional information about the structure of product development processes (PDP). These are supposed to map the flow of information and data as well as to provide a deeper understanding of the company's procedures. Process networks commonly represent dependencies of tasks and/or social contacts. Treating tasks as nodes in a network allows for a comparison of their position within the process. This way, it is possible to characterize certain actions according to their network attributes. In order to fully describe a PDP, it is, however, necessary to include other influencing factors as well. For example, there are only few approaches examining the impact of quality and progress on the process artefacts (such as CAD-files). The goal of this paper is to clarify what information is necessary to precisely describe a PDP in a network model. This covers a statement about the level of detail, general structure and dynamic of the networks.

A. Chahin (✉) · K. Paetzold
Institute of Technical Product Development, Universität der Bundeswehr München, Neubiberg, Germany
e-mail: abdo.chahin@unibw.de

J. Hoffmeister · V. Salehi
Department of Applied Sciences and Mechatronics, University of Applied Sciences Munich, Munich, Germany

G. Fanmuy et al. (eds.), *Complex Systems Design & Management*,
DOI 10.1007/978-3-319-49103-5_20

Challenges of Agile Development: A Cause-and-Effect Analysis

Tobias Sebastian Schmidt and Kristin Paetzold

Abstract Agile development as an alternative to traditional plan-driven approaches gains rising popularity in both software and non-software industries due to its advantages in dynamic and uncertain environments. Although its implementation challenges are widely explored, interdependencies between them are mostly neglected in recent papers. Practice and academia, therefore, often try to find local optimizations without (a) considering the interdependencies and (b) differentiating between causes and effects. By using the network theory this investigation sets up a directed network containing 241 challenges (nodes) and 360 dependencies (edges) and executes a cause-and-effect analysis. To identify challenges that are most crucial and, thus, of highest importance for future research, the analysis takes each challenges' (i) degree of being a cause, (ii) impact and (iii) range of influence into account. 'Granting freedom of action and decision', 'integrating agile methods in traditional organizations' and 'composing agile teams' turn out to be the top three challenges.

T.S. Schmidt (✉)
Institute for Innovation and Technology Management,
Werner-Heisenberg-Weg 39, 85577 Neubiberg, Germany
e-mail: tobias.schmidt@unibw.de

K. Paetzold
Institute for Technical Product Development,
Werner-Heisenberg-Weg 39, 85577 Neubiberg, Germany
e-mail: kristin.paetzold@unibw.de

G. Fanmuy et al. (eds.), *Complex Systems Design & Management*,
DOI 10.1007/978-3-319-49103-5_21

MBSE and MBSA with Capella and Safety Architect Tools

Marc Sango, Frédérique Vallée, Anne-Catherine Vié, Jean-Luc Voirin, Xavier Leroux and Véronique Normand

Abstract The development of critical systems is a challenging task that requires collaborative work for various purposes: specification, design and verification. Today, no single modeling language and environment covers all these aspects. ARCADIA and Capella© are Model-based System Engineering (MBSE) method and tool developed for the system design process. ARCADIA/Capella also adopts a viewpoint-based description to describe engineering specialty, such as the safety engineering. Safety Architect© is a MBSA (Model Based Safety Analysis) tool developed by ALL4TEC to analyze the robustness of design models. Indeed, Safety Architect can use design models imported from usual modelling tools, such as Capella, in order to perform classical safety analyses: automatic deduction of fault tree of the identified feared events. In this paper, we present our MBSE and MBSA approach developed in the Clarity project around Safety viewpoint in Capella and the import legacy into Safety Architect in order to realize safety analysis.

M. Sango (✉) · F. Vallée · A.-C. Vié
ALL4TEC, Changé, France
e-mail: Marc.Sango@all4tec.net

F. Vallée
e-mail: Frederique.Vallee@all4tec.net

A.-C. Vié
e-mail: Anne-Catherine.Vie@all4tec.net

J.-L. Voirin
Thales Mission Systems, Paris, France
e-mail: Jean-Luc.Voirin@all4tec.net

X. Leroux
Thales Global Services, Paris, France
e-mail: Xavier.Leroux@all4tec.net

V. Normand
Thales Research and Technology, Paris, France
e-mail: Veronique.Normand@all4tec.net

G. Fanmuy et al. (eds.), *Complex Systems Design & Management*,
DOI 10.1007/978-3-319-49103-5_22

Resilience Analysis on Infrastructure Networks with Heterogeneous Nodes

C.Y. Lam and K. Tai

Abstract The analysis of infrastructure network reliability is an important task required for disruption prevention, protection or recovery planning. In order to truly encapsulate the actual structure of infrastructure networks, it is proposed to analyze the infrastructure networks with heterogeneous nodes, i.e. the nodes with distinct operating features in the network. In this paper, an infrastructure network with heterogeneous nodes is modeled as a graph with a set of nodes with supply feature, a set of nodes with demand feature, and a set of connections between the nodes. The network resilience can then be evaluated by the weighted sum of all the resilience of the demand nodes, so the proposed resilience analysis approach can be used to indicate the ability of the network to resist disruption.

C.Y. Lam (✉)
Hiroshima University, 4-1, Kagamiyama 1-Chome, Higashi-Hiroshima 739-8527, Japan
e-mail: cylam@hiroshima-u.ac.jp

K. Tai
Nanyang Technological University, 50 Nanyang Avenue, Singapore 639798, Singapore

G. Fanmuy et al. (eds.), *Complex Systems Design & Management*,
DOI 10.1007/978-3-319-49103-5_23

Direct Democracy as the Keystone of a Smart City Governance as a Complex System

Claude Rochet

Abstract We analyze the "smartness" of the city as based on an organic evolving system. The smart city, in this perspective, is not a first order cybernetic self-regulating system which the number of parameters and variations could be finite which culd be modeled top-down by an engineer. On the contrary, the smart city is an autopoietic ecosystem (Maturana) and an adaptive system as promoted by the second order cybernetics, able to evolve as a dissipative system thanks to its internal interactions faced with the variations of its environment.

Consequently, from the perspective of the governance of such a system, the research question is "How to conceive the government of a system which doesn't need to be governed" being autopoietic and self regulated and self evolving?

(1) We rely on the literature on second order cybernetic (Joslyn and Heylighen) and to the critics of contemporaneous urban planning (Jane Jacobs, Lewis Mumford) based on the failure of centralized and deterministic planning (Urban planning in the USA, monocities in the former Soviet Union, Le Corbusier and Niemeyer in Europe and Latin America) to consider the smart city as a systemic *emergence* on an organic mode (Alexander, Mumford).

(2) We assume, basing on E. Oström works and research on self regulating and self sustainable human systems (Oström, Greif,) and Weick's perspective of "enactment" aiming at the transformation of a social construct, here applied to a smart city development project by its own actors, as a way to develop autopoietic properties in a governance system. that the key issue is for the stake holders to share a common goal and vision, that is to stay a share vision of the Common Good that will the heart of the functioning of the city. We use and test an approach based on design thinking and define a step by step methodology. Therefore, the problem turns into "conceiving the government

C. Rochet (✉)
LAREQUOI, Université de Versailles Saint-Quentin-en-Yveline, Versailles, France
e-mail: claude.rochet@univ-amu.fr

G. Fanmuy et al. (eds.), *Complex Systems Design & Management*,
DOI 10.1007/978-3-319-49103-5_24

of a man made system that doesn't need government since it has acquired autopoeitic properties".

(3) We draft a methodology to design the system as a whole, putting emphasis on the contribution of digital technologies and their interfaces with human system, and we conclude on the key features for a smart city governance.

We infer from this what would be the key competencies to be mastered by public managers and the stakeholders of the smart city, and conclude with the proposal of a theoretical and practical training program.

Fast and Extensive Model Based Project Plan Building in Nuclear Industry

Christian Marie, Gilles Beuzelin, Samuel Boutin and Eric Nicole

Abstract Areva investigated a Model Based approach for setting up the Work Breakdown Structure of a nuclear plant project performed in collaboration with other key industrial partners of the energy domain.

Problems to be solved included:

- Manage collaboration between several industrial partners having their own processes, methods and approaches.
- Generate the Work Breakdown Structure (WBS) and Work Packages Descriptions (WPD) for the project.

Modeling benefits were:

- Convergence on a set of generic processes to be applied together with associated standard document types: specifications, justification, design, validation… documents.
- Quick production of WBS from the Product Breakdown Structure (PBS) (e.g. we issued a PBS with ~100 items and got a WBS with more than 100 Work Packages and 1000 documents and items) as a basis for cost analysis and planning.
- An objective and neutral support for project plan enabling more efficient collaboration.

C. Marie
AREVA TA, Route de Saint Aubin, 91190 Villiers-le-Bâcle, France
e-mail: Christian.marie@areva.com

G. Beuzelin
AREVA, Tour AREVA, 1, place Jean Millier, 92 400 Courbevoie, France
e-mail: gilles.beuzelin@areva.com

S. Boutin · E. Nicole (✉)
Knowledge Inside, 7B rue Jean Mermoz, 78000 Versailles, France
e-mail: eric.nicole.bp@k-inside.com

S. Boutin
e-mail: samuel.boutin@k-inside.com

G. Fanmuy et al. (eds.), *Complex Systems Design & Management*,
DOI 10.1007/978-3-319-49103-5_25

B4B, a System of System Development Based on Systems Engineering Processes

Yann Chazal, Philippe Toussaint and Do-Hieu Trinh

arKItect SEA, a Systems Engineering (SE) modeler has been used by Renault and Bouygues Energies & Services to manage SE processes of a Batteries for Buildings (B4B) system. B4B is a concept reusing batteries of electric vehicles (second life), as a storage facility for energy management and renewables integration. The project started in 2012, was completely new from many viewpoints: new partnership with actors using different processes; innovative product and service offer connected and evolutive including safety concerns.

In order to overcome these challenges we have established a common SE model addressing the SE process: managing all SE data in a modeler, managing data consistency (requirements allocations, functional and system architecture), generating all specification documents toward developers and suppliers, enabling safety analysis faithfully with SE model.

Y. Chazal
Renault SAS, Saint Priest Cedex, France
e-mail: Yann.chazal@renault.com

P. Toussaint (✉)
Knowledge Inside, Versailles, France
e-mail: Philippe.toussaint@k-inside.com

D.-H. Trinh
Bouygues Energies et Services, Saint Quentin en Yvelines, France
e-mail: dh.trinh@bouygues-es.com

G. Fanmuy et al. (eds.), *Complex Systems Design & Management*,
DOI 10.1007/978-3-319-49103-5_26

Categorizing Technical Change in a System: Resolving Some of the Shortcomings in Henderson & Clarck's (1990) Framework

Mohammadreza Arasti and Mahdi Khaleghi

Abstract Henderson and Clarck (1990) have introduced the sole framework which classifies technical change in a system using two measures: degree of changes in components and intensity of changes in the linkage between components. Although this two dimensional framework is useful for understanding the congruence between different kinds of technical change, their consequences for the system's performance and their required capabilities, it ignores the vastness and relative importance of changes. To cope with this challenge, we propose adding a third dimension entitled "Change Magnitude" to their framework which contains a spectrum from changes in just one peripheral component or linkage to changes in all peripheral as well as core components and linkages. The resolved framework, presents an octal categorization of technical change in systems which provides a better basis for classification.

.

M. Arasti (✉)
Center for Research in Technology & Innovation Management in Complex, Industrial Systems (CRiTIMiX) at Graduate School of Management & Economics, Sharif University of Technology, Azadi Ave., Tehran, Iran
e-mail: arasti@sharif.edu

M. Khaleghi (✉)
Faculty of Management at University of Tehran & CRiTIMiX (Center for Research in Technology & Innovation Management in Complex Industrial Systems) at Sharif University of Technology, Next to Nasr Bridge, North Kargar St., Tehran, Iran
e-mail: Khaleghi.m@ut.ac.ir; khaleghimahdi@yahoo.com

G. Fanmuy et al. (eds.), *Complex Systems Design & Management*,
DOI 10.1007/978-3-319-49103-5_27

Exploring Early Stage Cost-Estimation Methods Using Off-the-Shelf Tools: A Preliminary Study

Haifeng Zhu, Narek Shougarian, Greg Ojard, Kaushik Sinha, Oliver de Weck and Eileen Arnold

Abstract Cost analysis is challenging for multiple reasons, one of which is the lack of historical data due to proprietary issues, or significant work required to make it useful for a particular application and domain of interest. In addition, to support system engineering methods such as Design Space Exploration, both component- and engine-level costs must be supported. This paper presents the results of a preliminary study on a tool that can be used to estimate the development cost for a set of airplane-engine architecture models using publicly available off-the-shelf tools. Our tool focuses on supporting complex system engineering tool chains and methods that require strong interoperability with different tools in a networked environment. The tool, through its architecture, allows the inclusion of supports for early stage cost analysis without directly using historical data, and both system- and component-level cost generations. We describe our approach, tools, estimation process and possible use cases.

Keywords Cost-estimation · Early stage

This research was partially developed with funding from the Defense Advanced Research Projects Agency (DARPA) and the Air Force Research Laboratory (AFRL). The views, opinions, and/or findings expressed are those of the author(s) and should not be interpreted as representing the official views or policies of the Department of Defense or the U.S. Government. Distribution Statement "A" (Approved for Public Release, Distribution Unlimited).

H. Zhu (✉) · G. Ojard
United Technologies Research Center, 411 Silver Lane, East Hartford, CT 06018, USA
e-mail: zhuhf@utrc.utc.com

N. Shougarian · K. Sinha · O. de Weck
Massachusetts Institute of Technology, 77 Massachusetts Ave. Building E38-532/33-410, Cambridge, MA 02139, USA

E. Arnold
UTC Aerospace Systems, 4747 Harrison Ave., Rockford, IL 61108, USA

G. Fanmuy et al. (eds.), *Complex Systems Design & Management*,
DOI 10.1007/978-3-319-49103-5_28

A Framework for Understanding the Complexity of Regional Production Networks: A Case Study

Larissa Statsenko and Vernon Ireland

Abstract A regional production network could be viewed as a complex network, consisting of an intertwined set of supply chains in a bounded geographical space, linking multiple customers in a particular industry with their associated suppliers. To avoid suboptimal decisions, supply chain managers and policy makers need to recognise the structural complexity of the regional production networks in which the individual supply chains are embedded. The authors propose a framework that allows for the identification of complexity traits in the regional production network structure, which provides an insight into its functionality and operational characteristics. The framework is based on the identification of network topology and structural parameters, including density, clustering and average path length. These parameters are indicative of network responsiveness, adaptability and resilience. The authors have applied the proposed framework to empirical data from the South Australian resource extraction sector to highlight how the regional production network structure could be used as a dashboard to assist both practitioners and policy makers in supply chain governance decision making.

L. Statsenko (✉) · V. Ireland
University of Adelaide, 10 Pulteney Street, Adelaide, SA, Australia
e-mail: larissa.statsenko@adelaide.edu.au

V. Ireland
e-mail: vernon.ireland@adelaide.edu.au

G. Fanmuy et al. (eds.), *Complex Systems Design & Management*,
DOI 10.1007/978-3-319-49103-5_29

Author Index

G. Fanmuy et al. (eds.), *Complex Systems Design & Management*,
DOI 10.1007/978-3-319-49103-5

MIX
Papier aus verantwortungsvollen Quellen
Paper from responsible sources
FSC® C105338

If you have any concerns about our products,
you can contact us on
ProductSafety@springernature.com

In case Publisher is established outside the EU,
the EU authorized representative is:
Springer Nature Customer Service Center GmbH
Europaplatz 3, 69115 Heidelberg, Germany

Printed by Libri Plureos GmbH
in Hamburg, Germany